T0135570

Dissertation zur Erlangung des Doktorgrades
der Fakultät für Chemie und Pharmazie
der Ludwig–Maximilians–Universität München

Magnetic properties of
transition metal surfaces
and GaAs/Fe heterogeneous systems

Michal Košuth

aus

Bratislava, Slowakische Republik

2007

Bibliografische Information der Deutschen Nationalbibliothek

Die Deutsche Nationalbibliothek verzeichnet diese Publikation in der
Deutschen Nationalbibliografie; detaillierte bibliografische Daten sind
im Internet über http://dnb.d-nb.de abrufbar.

ISBN 978-3-8325-1806-6

Logos Verlag Berlin GmbH
Comeniushof, Gubener Str. 47,
10243 Berlin
Tel.: +49 030 42 85 10 90
Fax: +49 030 42 85 10 92
INTERNET: http://www.logos-verlag.de

Contents

1 Introduction   **9**

2 Theoretical Framework   **13**
  2.1 Density Functional Theory 13
     2.1.1 The Hohenberg-Kohn theorems 14
     2.1.2 Relativistic Density Functional Theory 17
  2.2 Korringa-Kohn-Rostoker (KKR) method 20
     2.2.1 Single particle Green's function 21
     2.2.2 Multiple scattering theory 23
     2.2.3 Solution of the single-site problem –
          Fully-relativistic treatment 26
     2.2.4 Treatment of disordered alloys 32
  2.3 Tight-binding KKR (TB-KKR) method 33
     2.3.1 Transformation to the tight-binding form 34
     2.3.2 Transformation to an arbitrary reference system . 35
     2.3.3 Tight-binding KKR method for slab geometry . . 36
     2.3.4 N-scaling procedure 38
  2.4 Magnetic anisotropy energy (MAE) 40
     2.4.1 Magnetocrystalline anisotropy energy 41
     2.4.2 Shape anisotropy 42
     2.4.3 Bruno's and van der Laan's formulas 42

3 Fe surfaces and clusters   **45**
  3.1 Introduction 45
  3.2 Fe surfaces 47
     3.2.1 Structure and computational details 47
     3.2.2 Densities of states 47
     3.2.3 Magnetic properties 49

	3.2.4	Magnetic anisotropy	53
3.3	Fe clusters		53
	3.3.1	Structure and computational details	53
	3.3.2	Magnetic properties	55
3.4	From clusters to surfaces		60
	3.4.1	DOS comparison	60
	3.4.2	Comparison between magnetization profiles of clusters and surfaces	63
	3.4.3	Systematic trends in magnetic moments	66

4 FePt surfaces — **73**
4.1	Introduction		73
4.2	Structure and computational details		74
4.3	Electronic properties		76
	4.3.1	Bulk FePt	76
	4.3.2	Fe- and Pt-terminated FePt(001) surfaces	76
	4.3.3	FePt(100) and disordered $Fe_{0.50}Pt_{0.50}(001)$ surface	78
4.4	Magnetic properties		78
	4.4.1	Magnetic moments	78
	4.4.2	Magnetic anisotropy	81

5 GaAs/Fe heterogeneous systems — **85**
5.1	Introduction		85
5.2	Studied systems		87
5.3	GaAs/Fe multilayers		88
	5.3.1	Spin magnetization profile	88
	5.3.2	XMCD investigations	91
5.4	Fe films on GaAs (GaAs/nFe systems, n=1–7)		95
	5.4.1	Spin and orbital magnetization profile	95
	5.4.2	Validity of Bruno's and van der Laan's formulas	96
	5.4.3	Out-of-plane magnetic anisotropy	97
	5.4.4	In-plane magnetic anisotropy	98
5.5	Influence of Au covering layers		101
	5.5.1	Introduction	101
	5.5.2	GaAs/nFe/3 fcc Au systems, n=1–7	102
	5.5.3	GaAs/4(7)Fe/bcc Au systems	106
5.6	Co marker in Fe film on GaAs		114

5.6.1 Introduction 114
5.6.2 Studied systems 115
5.6.3 XMCD sum rules 116
5.6.4 Spin and orbital magnetization profiles 118
5.6.5 Co marker detecting the Fe magnetization profile 120

6 Summary   123

Bibliography   127

Acknowledgements   145

Curriculum vitae – Lebenslauf   147

List of Publications (1999-2006)   149

Chapter 1

Introduction

Phenomena such as magnetocrystalline anisotropy, magnetooptical effects, magnetic circular dichroism etc. caused by strong coupling among spin and orbital degrees of freedom are central issues in the physics of transition metal compounds over the last decade. One of the aims of the material science today is to design magnetic storage devices of smaller size, microchips on a magnetic basis exploiting, besides the charge, also the spin degree of freedom. Theoretical solid state physics supports this technological progress. The interesting task of the theory lies in explaining the experimental observations, the more thrilling task is the investigation of the properties of novel devices and the design of these devices itself. As the calculations are less cost and time demanding than the experiments, they can be successfully used for modifying already existing devices.

In Chap. 2 the fundamentals of density functional theory (DFT) will be shown. The DFT represents the basis for the self-consistent band structure calculations performed in this work. The Korringa-Kohn-Rostoker (KKR) method to solve the Kohn-Sham equations, based on the multiple scattering formalism, will be discussed in detail. The use of the tight-binding version of the KKR method allows one to investigate layered systems of bigger sizes than it was possible with the standard KKR method. For the proper functioning of a magnetic device the direction of the spontaneous magnetization (the easy axis) and the energy required to alter the magnetization direction, the so-called magnetic anisotropy energy (MAE), has to be known. The MAE can be obtained as a difference of the total energies of the system for two different magnetization directions. Van Vleck [1] suggested that the origin of this anisotropy is the interaction of the magnetization with the crystal lattice, this implies it has its origin in the spin-orbit coupling. As the spin-orbit coupling is of relativistic origin, the fully-relativistic treatment to the tight-binding KKR (TB-KKR) method is used in this thesis.

For some technological applications, an in-plane magnetization is desirable. For example, in-plane anisotropy is useful for longitudinal recording, magnetostrictive and inductive heads and media for magnetic-field sensors (where a small anisotropy field is desired). Generally, if a ferromagnetic material lacks a high degree of magnetocrystalline anisotropy in the bulk (e.g., bcc Fe), the moments will lie in the layer (film) plane so as to minimize the free energy of the system (i.e., the magnetization direction is determined by shape anisotropy). MAE in bulk cubic $3d$ transition metals is a very small quantity of only a few μRy/atom [2, 3] that becomes enhanced by two orders of magnitude in the surface region of the transition metal systems. In Chap. 3 the magnetic properties and the MAE of Fe surface systems will be discussed.

Magnetic layers with strong out-of-plane anisotropy, on the other hand, are of great interest for perpendicular recording media with predicted performances of up to 1 Tbit/in^2, in comparison with present commercial hard disks that have storage densities around 100 Gbit/in^2. Hence, the potential need for materials with higher out-of-plane magnetocrystalline anisotropy than the current conventional cobalt alloys for future high-density recording media is obvious [4]. One of the hot candidates is FePt because of its high out-of-plane MAE. The influence of the surface and of the disorder on the magnetic anisotropy of FePt systems will be discussed in Chap. 4.

Spin- or magnetoelectronics, that exploits besides the charge also the spin of the electrons as an additional degree of freedom, receives a lot of interest recently because of its great technological potential [5]. A prominent example is the tunneling magnetoresistance (TMR) that denotes the observation that the tunneling current across a tunneling barrier that separates two magnetic layers depends on the relative orientation of the magnetization of these layers. A firm basis for the understanding of the transport properties of the Fe/GaAs/Fe systems cannot be achieved without a full understanding of their electronic and magnetic properties. This route is followed in Sec. 5.3 by studying GaAs/Fe multilayer systems using standard KKR band structure techniques. Because the electronic and magnetic properties at the interface of GaAs/Fe systems are of central importance for the transport properties, special emphasis will be laid on these. In particular, the spin magnetization profiles will be presented. Making use of the Coherent Potential Approximation (CPA), the influence of interdiffusion at the GaAs–Fe interface within the multilayer model will be investigated in addition. To stimulate corresponding experimental work the X-ray magnetic circular dichroism (XMCD) will be shown for the $L_{2,3}$-edges absorption spectra of Ga as well as As.

The magnetic anisotropy of thin Fe films on GaAs will be discussed in Sec. 5.4, namely because of its central importance for the operation of the Fe/GaAs/Fe TMR-system. Whereas in the bulk bcc Fe a fourfold biaxial in-plane anisotropy is developed, after depositing Fe on a GaAs substrate (having the twofold symmetry of the zinc-blende structure) an additional twofold uniaxial in-plane contribution appears. In line with experimental findings an in-plane anisotropy with the easy axis pointing along the [110]-direction could be found in our calculations. The theoretical results for the anisotropy energy were reproduced in a semi-quantitative way by the models of Bruno and van der Laan, respectively, which relate the anisotropy energy to the anisotropy of the spin-orbit induced orbital magnetic moment.

In most experiments, the surface of Fe on GaAs is covered by protective layers of a noble metal (in most cases Au) to avoid the corrosion of Fe. Independent on the type of the covering layer, all experiments show a strong uniaxial (twofold) in-plane anisotropy with the easy axis in the [110]-direction and the hard axis in the [1$\bar{1}$0]-direction. In Sec. 5.5 we will concentrate on the influence of Au covering layers on the magnetic anisotropy of Fe films on GaAs. In addition, the layer by layer decomposition of the MAE will give a clearer view on the issue.

In order to examine the layer-resolved magnetic properties of the Fe film on GaAs substrate in an experiment, Giovanelli et al. [6–8] introduced a Co marker in the Fe film. The X-ray magnetic circular dichroism (XMCD) sum rules allow the connection of the measured XMCD spectra with the spin and orbital moments. Theoretical investigation of the spin and orbital magnetization of the Co marker in Fe film on GaAs and the comparison with the experimental results will be the topic of Sec. 5.6. In addition, the question of the feasibility of the Co marker in detecting the Fe magnetization profile will be discussed.

The work concludes with a summary in Chap. 6.

Chapter 2

Theoretical Framework

2.1 Density Functional Theory

The last forty years have borne witness to spectacular advances, due to the development of both theoretical techniques and computer technology, in the ability to describe the physical and electronic structure of condensed materials using self-consistent first-principle quantum mechanical methods. Many of the techniques that have been developed are based on density functional theory (DFT) and the aim of this chapter is to describe the basis of this theory.

In a system of interacting electrons the electrons do not move independently of each other, i.e. their motion is correlated. This means that the ground state wave function of this system cannot be expressed simply in terms of individual wave functions of non-interacting electrons as the interacting system depends on the motion of all electrons. Consequently, the problem quickly becomes very complicated and is impossible to solve exactly even for a simple system of two electrons.

Fortunately, although the electron-electron interaction within the material is strong, the exchange-correlation effects are relatively weak and so it is possible to develop a method that describes the electrons in the system as if they moved independently on each other. Such a method is DFT which differs from Hartree-Fock theory as it uses electron densities, rather than wave functions, as the basic quantity to describe the properties of a material.

DFT was first introduced by Hohenberg and Kohn [9] in the early 1960s. To obtain the ground state properties of a solid, complex many body problem is shrinked to a considerably simpler single particle problem. All electron-electron interactions are represented by an effective potential in which a single particle is moving. The effective potential contains the Coulomb interaction with all other particles, as well as quantum mechanical exchange and correlation interactions.

As exchange-correlation interactions cannot be described exactly in the one particle system, one has to choose an applicable and convenient approximation to the exchange-correlation potential. Up to now, the most widely used one is the local density approximation (LDA). It is applicable for systems with slowly varying or high electron densities. Theoretically calculated properties of solids (densities of states, magnetic moments and electronic energies) can be calculated in this way with a relatively small computational effort.

Our theoretical investigations will be based on the local spin density approximation (LSDA) scheme to deal with exchange and correlation effects. The Vosko, Wilk, and Nusair parametrization of the exchange-correlation potential [10] has been used throughout the whole thesis. The reliance on LSDA as opposed to the generalized gradient approximation (GGA) is justified in our study because we focus on magnetic properties of fixed-geometry systems. Although GGA was found to be superior to LSDA in exploring structural properties of transition metals [11], its benefit in magnetic studies is still questionable [12–14].

2.1.1 The Hohenberg-Kohn theorems

The formulation of DFT started with the proof of Hohenberg and Kohn [9] theorems in 1964. They have shown that the ground state density $n(\vec{r})$ of a N-electron system uniquely determines the ground state. Using these theorems, instead of the complex ground state wave function the many body problem is described by the simpler ground state density.

Within DFT the N interacting electrons are determined by an external potential $V_{\text{ext}}(\vec{r})$ and is shown that in turn the ground state density $n(\vec{r})$ uniquely determines the total potential $V(\vec{r})$. From Hohenberg-Kohn theorems it is clear that one may write the total ground state energy as an unique functional of the density. The second statement is the variational principle introducing the minimization condition: the energy corresponding to any density $n(\vec{r})$ must be greater than the energy of the ground state i.e.

$$E[n(\vec{r})] > E_{GS}[n_{GS}(\vec{r})] \ , \tag{2.1}$$

where E_{GS} and $n_{GS}(\vec{r})$ are the ground state energy and density respectively. The main idea of DFT is to describe a many body system as a simple non-interacting system having the same density like the many body system. The non-interacting system can be characterized by the single particle Hamiltonian $H_0 = T_0 + V_{\text{eff}}$. To construct the effective averaged potential $V_{\text{eff}}(\vec{r})$ of a non-interacting system, the energy functional $E[n(\vec{r})]$ of the many body system will be minimized. This functional was split by Kohn and Sham [15] into its various contributions:

$$\begin{aligned} E[n(\vec{r})] &= T_0[n(\vec{r})] + \iint \frac{n(\vec{r})n(\vec{r}')}{|\vec{r} - \vec{r}'|} \mathrm{d}^3 r \mathrm{d}^3 r' + E_{\text{xc}}[n(\vec{r})] \\ &\quad + \int V_{\text{ext}}(\vec{r})n(\vec{r})\mathrm{d}^3 r \ , \end{aligned} \tag{2.2}$$

where $T_0[n(\vec{r})]$ is the kinetic energy of a system of N non-interacting electrons, the second term is the classical Coulomb potential and $E_{\text{xc}}[n(\vec{r})]$ is the exchange-correlation energy. The exchange-correlation energy E_{xc} is defined as the difference between the exact energy and other contributions that may be exactly evaluated numerically. The last term describes the energy due to the interaction of electrons with the external potential $V_{\text{ext}}(\vec{r})$. For the only unknown, the exchange-correlation energy $E_{\text{xc}}[n(\vec{r})]$, an approximation has to be made.

The Kohn-Sham equations appear as the Euler-Lagrange equations of the variational principle

$$\frac{\delta}{\delta \psi_i^*(\vec{r})} E[n(\vec{r})] = 0, \qquad i = 1, ..., N \tag{2.3}$$

under the condition that the number of electrons $N = \int n(\vec{r})\mathrm{d}^3 r$ in the system is conserved:

$$\left[-\nabla^2 + \left(V_{\text{ext}}(\vec{r}) + \int \frac{n(\vec{r}')}{|\vec{r} - \vec{r}'|} \mathrm{d}^3 r' + \frac{\delta E_{\text{xc}}[n(\vec{r})]}{\delta n(\vec{r})} \right) \right] \psi_i(\vec{r}) = \epsilon_i \psi_i(\vec{r}) \tag{2.4}$$

The energies ϵ_i and the wave functions ψ_i are known as Kohn-Sham energies and Kohn-Sham wave functions respectively. It must be noted, however, that these are in a strict sense just auxiliary quantities that have no physical meaning. Anyway, the energies ϵ_i are often physically interpreted as excitation energies. Solving the Kohn-Sham equations in a self-consistent iterative way, one gets the ground state density and the total energy. The part in round brackets in Eq. (2.4) is interpreted as the effective single particle potential $V_{\text{eff}}(\vec{r})$.

As mentioned above, the crucial problem in the presentation of DFT and in the solution of Kohn-Sham equations is the expression for the exchange-correlation energy. In practice one always has to introduce an approximation of this energy since its form is not explicitly known. Up to now, the most widely used one is the local density approximation (LDA), which is used for the exchange-correlation energy $E_{\text{xc}}[n(\vec{r})]$:

$$E_{\text{xc}}[n(\vec{r})] = \int n(\vec{r})\epsilon_{\text{xc}}[n(\vec{r})]\mathrm{d}^3r \ , \qquad (2.5)$$

where $\epsilon_{\text{xc}}[n(\vec{r})]$ is the exchange-correlation energy per electron of the homogeneous electron gas with density $n(\vec{r})$ [16]. The LDA is exact by definition in two limiting cases: for slowly varying or high electron densities. DFT with the LDA is a highly powerful tool for the description of a vast range of material properties and obtained results show good agreement with experiment.

To be able to discuss spin-polarized magnetic systems, one has to include spin in the DFT [17]. Here one considers a spin-polarized N-electron system described by Hamiltonian of the form

$$H = T(\vec{r}) + U(\vec{r}, \vec{r}') + \int \left[V_{\text{ext}}(\vec{r})n(\vec{r}) + \vec{B}(\vec{r})\vec{m}(\vec{r})\right]\mathrm{d}^3r \ , \qquad (2.6)$$

where T and U describe kinetic energy and electron-electron interaction which are defined in the same way like in the nonmagnetic case. The potential energy includes now an extra term as a result of the coupling between the electron spin and an external magnetic field.

If the external magnetic field $\vec{B}(\vec{r})$ and the quantization axis point along the [001]-direction, the densities $n(\vec{r})$ and $m(\vec{r})$ are related to the densities $n^{\uparrow}(\vec{r})$ and $n^{\downarrow}(\vec{r})$ of the spin-$\uparrow$ and spin-$\downarrow$ electrons, respectively, via the equations

$$\begin{aligned}
n(\vec{r}) &= n^{\uparrow}(\vec{r}) + n^{\downarrow}(\vec{r}) & (2.7) \\
m(\vec{r}) &= n^{\uparrow}(\vec{r}) - n^{\downarrow}(\vec{r}) \ . & (2.8)
\end{aligned}$$

Then the Kohn-Sham equations (2.4) can be generalized within spin density functional theory (SDFT) to the form

$$\left\{\left[-\nabla^2 + V_{\text{eff}}(\vec{r})\right] I_2 - \mu_{\text{B}}\sigma_z B_{\text{eff}}(\vec{r})\right\}\psi_i(\vec{r}) = \epsilon_i\psi_i(\vec{r}) \ , \qquad (2.9)$$

where I_2 is the 2×2 identity matrix, σ_z is the following Pauli spin matrix [18]

$$\sigma_z = \begin{pmatrix} 1 & 0 \\ 0 & -1 \end{pmatrix} \ , \qquad (2.10)$$

while $\psi_i(\vec{r})$ is a spinor with components $\psi_{i\uparrow}(\vec{r})$ and $\psi_{i\downarrow}(\vec{r})$ with the effective fields being defined by

$$V_{\text{eff}}[n, m](\vec{r}) \ = \ V_{\text{ext}}(\vec{r}) + \int \frac{n(\vec{r}\,')}{|\vec{r} - \vec{r}\,'|}\mathrm{d}^3r\,' + \frac{\delta F_{\text{xc}}[n, m]}{\delta n(\vec{r})} \qquad (2.11)$$

$$B_{\text{eff}}[n, m](\vec{r}) \ = \ B_{\text{ext}}(\vec{r}) + \frac{\delta E_{\text{xc}}[n, m]}{\delta m(\vec{r})} \ . \qquad (2.12)$$

2.1.2 Relativistic Density Functional Theory

Relativistic effects, in particular the spin-orbit coupling, give rise to a great number of interesting and technologically important phenomena in magnetic systems. Amongst others, the magnetic circular dichroism and the magnetic anisotropy will be discussed in this thesis. Relativistic effects can be included through various correction terms, as for example the inclusion of the rest mass or spin-orbit coupling terms. Nevertheless, a fully-relativistic treatment of DFT has been developed. This implies the use of four component basis functions derived from the proper Dirac equation for a magnetic solid. The first step was made by Rajagopal and Callaway [19], leading in principle to a current density formalism (CDFT) with the expectation value of four-current density as the central quantity. However, because of the many unsolved problems connected with this general scheme a relativistic version of spin density functional theory has been suggested by MacDonald and Vosko [20], as well as by Ramana and Rajagopal [21].

The interaction of the external magnetic field $\vec{B}_{\text{ext}}(\vec{r}) = \vec{\nabla} \times \vec{A}_{\text{ext}}(\vec{r})$ with the electronic system is included in the Hamiltonian through the coupling of the Dirac current to the vector potential $\vec{A}_{\text{ext}}$:

$$H = T(\vec{r}) + U(\vec{r}, \vec{r}') + \int J^\mu(\vec{r}) A^\mu_{\text{ext}}(\vec{r}) \mathrm{d}^3 r \ . \tag{2.13}$$

Here the Dirac electronic system is described using the four-vector notation. The scalar and the vector potential are then combined to get the following expression for the external potential:

$$A^\mu_{\text{ext}}(\vec{r}) = \left(-\frac{1}{e} V_{\text{ext}}(\vec{r}), \vec{A}_{\text{ext}}(\vec{r}) \right) \ . \tag{2.14}$$

The electronic four-current density distribution has the form:

$$J^\mu(\vec{r}) = \left(-\frac{ec}{2} n(\vec{r}), \vec{J}(\vec{r}) \right) \ . \tag{2.15}$$

Assuming that the ground state of a noninteracting system can be described by a Slater determinant made up of N single particle wave functions ψ_i, the Dirac-Kohn-Sham equations can be written as:

$$\left[c\vec{\alpha} \cdot \left(-i\hbar\vec{\nabla} + \frac{e}{c}\vec{A}_{\text{eff}}(\vec{r}) \right) + \beta mc^2 + V_{\text{eff}}(\vec{r}) \right] \psi_i(\vec{r}) = \epsilon_i \psi_i(\vec{r}) \ , \tag{2.16}$$

with ϵ_i the eigenvalues, $\vec{A}_{\text{eff}} = \vec{A}_{\text{ext}} + \vec{A}_{\text{H}} + \vec{A}_{\text{xc}}$ the effective vector potential with the Hartree and exchange-correlation parts denoted by the subscripts H and xc. The 4×4 matrices α_i and β are Dirac matrices defined as [18]:

$$\alpha_i = \begin{pmatrix} 0 & \sigma_i \\ \sigma_i & 0 \end{pmatrix} \ , \tag{2.17}$$

$$\beta = \begin{pmatrix} I_2 & 0 \\ 0 & -I_2 \end{pmatrix} \ , \tag{2.18}$$

with σ_i being the standard Pauli matrices. Performing the variation of the total energy functional with respect to the four-current density $J^\mu(\vec{r})$, the following equations for the effective scalar and vector potentials are obtained:

$$V_{\text{eff}}(\vec{r}) = V_{\text{ext}}(\vec{r}) + \int \frac{n(\vec{r}')}{|\vec{r} - \vec{r}'|} \mathrm{d}^3 r' + \frac{\delta E_{\text{xc}}[\vec{J}^\mu(\vec{r})]}{\delta n(\vec{r})} \tag{2.19}$$

$$\vec{A}_{\text{eff}}(\vec{r}) = \vec{A}_{\text{ext}}(\vec{r}) + \frac{1}{c} \int \frac{\vec{J}(\vec{r}')}{|\vec{r} - \vec{r}'|} \mathrm{d}^3 r' + c\frac{\delta E_{\text{xc}}[\vec{J}^\mu(\vec{r})]}{\delta \vec{J}(\vec{r})} \ . \tag{2.20}$$

In Eqs. (2.19) and (2.20) the effective scalar $V_{\text{eff}}(\vec{r})$ and vector potentials $\vec{A}_{\text{eff}}(\vec{r})$, respectively, contain external contributions as first terms. The second term in Eq. (2.19) is the familiar Hartree potential, whereas in Eq. (2.20) it is the potential due to the Breit-interaction. Finally, the third terms are caused by exchange and correlation. Analogously to the non-relativistic DFT, the whole many body problem is sticked in the exchange-correlation potentials $\frac{\delta E_{\text{xc}}[\vec{J}^{\mu}(\vec{r})]}{\delta J^{\nu}(\vec{r})}$.

The exchange-correlation energy poses the most severe problem as this term includes the current-current interactions. The relativistic exchange-correlation functional $E_{\text{xc}}[\vec{J}^{\mu}(\vec{r})]$ has to include all the magnetic exchange-correlation effects, which are intrinsically relativistic in nature, like the retardation of the Coulomb interaction between electrons and the magnetic interaction between moving electrons, through its dependence on the spatial components of current. As there is no local approximation to $E_{\text{xc}}[\vec{J}^{\mu}(\vec{r})]$, using CDFT explicitly in practical calculations is complicated. An approximate relativistic version of SDFT has been workout by several authors [19, 20, 22]. The first step in this direction is the Gordon decomposition of the current density into its spin and orbital parts [18]. For moderate external magnetic fields, the orbital part can be neglected. The coupling of the spin part to the vector potential $\vec{A}_{\text{eff}}$ may alternatively be described by introducing the corresponding magnetization density

$$\vec{m}(\vec{r}) = \sum_i \psi_i^{\dagger}(\vec{r})\beta\vec{\sigma}\psi_i(\vec{r})\Theta(\mu - \epsilon_i) \ , \qquad (2.21)$$

and a coupling term analogous to Eq. (2.9)

$$\mu_{\text{B}}\vec{m}(\vec{r}) \cdot \vec{B}_{\text{eff}}(\vec{r}) \ . \qquad (2.22)$$

Finally, Eq. (2.16) can be rewritten completely analogous to give the relativistic Dirac-Kohn-Sham equation:

$$\left[-ic\hbar\vec{\alpha}\vec{\nabla} + \beta mc^2 + V_{\text{eff}}(\vec{r}) + \mu_{\text{B}}\beta\sigma_{\text{z}}B_{\text{eff}}(\vec{r})\right]\psi_i(\vec{r}) = \epsilon_i\psi_i(\vec{r}) \ , \qquad (2.23)$$

where a possible external magnetic field and the magnetization density pointing in the z-direction were assumed.

2.2   Korringa-Kohn-Rostoker (KKR) method

By means of the density functional theory the many body electronic problem has been reduced to the problem of a single electron moving independently in an effective potential created by the nuclei and all other electrons within the system. Instead of computing the Kohn-Sham orbitals directly, one can construct the Green's function corresponding to the Kohn-Sham Hamiltonian. To calculate all single particle expectation values of a crystal like charge and spin densities it is enough to know the Green's function. Thus it is possible to determine various electronic, magnetic and spectroscopic properties in a very elegant way. The KKR Green's function method has been developed by Dupree [23], Beeby [24] and Holzwarth [25] on the basis of a multiple scattering method proposed by Korringa [26], Kohn and Rostoker [27]. In the last three decades further developments for the KKR theory were undertaken, among others by Faulkner [28, 29], Faulkner and Stocks [30], Györffy and Stocks [31] for the non–relativistic case, and by Onodera and Okazaki [32], Strange *et al.* [33], Weinberger [34], Strange [18] and Ebert [35] for the relativistic one.

Among the outstanding features of the KKR method one has to emphasize that the Green's function transparently relates single-site scattering and geometrical or structural quantities, and that there are no preconditions with regard to the arrangement of the atoms, as e.g. spatial periodicity. Therefore the approach can be separated into two parts, the first supplies the solution corresponding to single-site scattering, while the second is dependent solely on the geometry of the system. Hence the KKR Green's function method is predestinated for the treatment of defect systems like vacancies and impurities [36], but also surfaces and interfaces [37]. A further important field of application are disordered alloys, for which the KKR Green's function method in conjunction with the Coherent Potential Approximation (CPA) [38] constitutes the best available single-site method of calculating the electronic structure [39].

The basic idea of KKR is that an incident wave coming to any given site is a superposition of the outgoing ones from all other sites. In order to determine the Green's function of the system at fixed energy, the first task is to determine the structure constants and the so-called t-matrix.

The structure constants depend on the crystal structure only and the t-matrix describes the scattering on each individual atomic scatterer characterized by non-overlapping potentials. Further, the construction of the so-called T-matrix in order to reproduce the scattering in the whole crystal is needed. Using the appropriate Dyson equation, the atomic t-matrices and the structure constants can be combined to the so-called scattering path operator τ. This τ describes all possible scattering events of a single electron on its way between two individual scattering centers. Thus it is a central quantity to construct the Green's function for the whole system.

In the next chapter, the single particle Green's operator will be introduced and its properties with regard to the KKR method will be given. As a common approximation the potential is regarded to be spherically symmetric, which leads to a significant reduction of computing time. Following this formalism the Green's function of the multiple scattering problem is set up in Sec. 2.2.2. Many of the calculated properties presented in this thesis (orbital magnetic moment, magnetic anisotropy energy) are of relativistic origin and their origin lies in the spin-orbit coupling. Therefore a fully-relativistic treatment of the KKR method is needed and will be described in Sec. 2.2.3.

Initially, the KKR method was developed for the treatment of infinite extended crystals. In this case the multiple scattering problem is solved by using the Fourier transformation and Brillouin zone integration in three dimensions. In the following, this method will be called the "standard" KKR method. However, by changing the boundary conditions to that of a two-dimensional system, one can also apply the KKR method to surfaces, multilayers, heterogeneous layered systems, etc. As layered systems investigated in this thesis obviously need this new approach, the Sec. 2.3 is devoted to the two-dimensional screened tight-binding KKR method.

2.2.1   Single particle Green's function

Let's consider the following eigenvalue problem, where $H(\vec{r})$ is a general Hermitian operator:

$$H(\vec{r})\phi_j(\vec{r}) = E_j\phi_j(\vec{r}) \qquad (2.24)$$

The Green's function associated to this Schrödinger equation is defined by

$$[E - H(\vec{r})]\, G(\vec{r}, \vec{r}\,', E) = \delta(\vec{r} - \vec{r}\,') \ , \tag{2.25}$$

with the given boundary conditions.

Using a complete set of orthonormal wave functions ϕ_j, e.g. $\sum_j \phi_j(\vec{r}\,')\phi_j^\dagger(\vec{r}) = \delta(\vec{r} - \vec{r}\,')$, the corresponding spectral representation of the Green's function may be written as

$$G(\vec{r}, \vec{r}\,', E) = \sum_j \frac{\phi_j(\vec{r}\,')\phi_j^\dagger(\vec{r})}{E - E_j} \ . \tag{2.26}$$

In the case of a continuous eigenvalues spectrum of a solid the summation in Eq. (2.26) must be replaced by an integral. This integral is not defined at real energies because of the singularities at E_j, therefore one has to include a small imaginary part to the energy. For this purpose the retarded $(+)$ and advanced $(-)$ Green's function

$$G^\pm(\vec{r}, \vec{r}\,', E) = \lim_{\epsilon \to 0^+} \sum_j \frac{\phi_j(\vec{r}\,')\phi_j^\dagger(\vec{r})}{E - E_j \pm i\epsilon} \tag{2.27}$$

is defined. We restrict ourselves to the retarded Green's function from now on and will use the notation G.

The Green's function can be used to calculate all observables, for example the density of states. Using the following identity for the Dirac delta function [40]:

$$\lim_{\epsilon \to 0^+} \frac{1}{E - E_j + i\epsilon} = P\left(\frac{1}{E - E_j}\right) - i\pi\delta(E - E_j) \tag{2.28}$$

one can show that

$$-\frac{1}{\pi}\,\mathrm{Im}\int \mathrm{Tr}\, G(\vec{r}, \vec{r}, E)\mathrm{d}^3r = \sum_j \delta(E - E_j) \tag{2.29}$$

where the quantity $\sum_j \delta(E - E_j)$ gives the number of states $n(E)$ at energy E_j. $n(E)$ is called the density of states (DOS) as $n(E)\mathrm{d}E$ gives the number of states for the chosen energy interval $\mathrm{d}E$. The expectation values of all other observables are obtained then by taking the trace of the product between the Green's function and the operator of interest:

$$\langle A \rangle = -\frac{1}{\pi}\,\mathrm{Im}\,\mathrm{Tr}\int^{E_\mathrm{F}} \mathrm{d}E \int \mathrm{d}^3r\, \hat{A}G(\vec{r}, \vec{r}, E) \ . \tag{2.30}$$

Splitting H into a Hamiltonian of an unperturbed reference system H_0 and a perturbation V so that $H = H_0 + V$, one can introduce the Green's function G_0 for the unperturbed system fulfilling the equation

$$[E - H_0(\vec{r})]\, G_0(\vec{r}, \vec{r}\,', E) = \delta(\vec{r} - \vec{r}\,') \ . \tag{2.31}$$

Considering the two Green's functions G and G_0 of the perturbed and unperturbed system, respectively, they are coupled by

$$\begin{aligned} G(\vec{r}, \vec{r}\,', E) = &\ G_0(\vec{r}, \vec{r}\,', E) \\ &+ \int G_0(\vec{r}, \vec{r}\,'', E) V(\vec{r}\,'') G(\vec{r}\,'', \vec{r}\,', E) \mathrm{d}^3 r'' \ . \end{aligned} \tag{2.32}$$

This is the so-called Dyson equation giving G in terms of self-consistent relation. Analogously the solution of the Schrödinger equation $\left[\hat{H}_0(\vec{r}) + V(\vec{r})\right] \psi(\vec{r}) = E\psi(\vec{r})$ can be written

$$\psi(\vec{r}) = \phi(\vec{r}) + \int G_0(\vec{r}, \vec{r}\,', E) V(\vec{r}\,') \psi(\vec{r}\,') \mathrm{d}^3 r' \ , \tag{2.33}$$

which is known as the Lippmann–Schwinger equation.

2.2.2 Multiple scattering theory

In order to eliminate the full solution $\psi(\vec{r})$ from the right hand side of the Lippmann–Schwinger equation (2.33), a scattering operator T can be introduced (for sake of simplicity, the Lippmann–Schwinger equation will be rewritten in operator form, with the coordinate and energy arguments dropped)

$$V\,|\,\psi\,\rangle = T\,|\,\phi\,\rangle \ . \tag{2.34}$$

Eq. (2.33) can be rewritten in terms of the free particle wave function:

$$|\,\psi\,\rangle = |\,\phi\,\rangle + G_0 T\,|\,\phi\,\rangle \ . \tag{2.35}$$

A similar form can be used for the Dyson equation (2.32)

$$G = G_0 + G_0 T G_0 \ . \tag{2.36}$$

Consequently, the scattering operator T describing all possible scatterings in the system, can be rewritten in the form:

$$T = V(1 + G_0 T) \ . \tag{2.37}$$

The effective potential in a solid can be subdivided into single non-overlapping potentials located at the atomic sites $\vec{R}_i$:

$$V(\vec{r}) = \sum_i V_i(\vec{r} - \vec{R}_i) \ . \tag{2.38}$$

Using Eq. (2.38) in Eq. (2.37) and defining the single-site scattering operator t^i corresponding to site i

$$t^i = V_i \left(1 + G_0 \, t^i\right) \ , \tag{2.39}$$

or, equivalently,

$$t^i = \left[1 - V_i \, G_0\right]^{-1} V_i \ , \tag{2.40}$$

we arrive at the following expansion for the T-operator of the crystal potential in terms of the single-site scattering operator t^i:

$$T = \sum_i t^i + \sum_{j \neq i} t^i \, G_0^{ij} \, T \ . \tag{2.41}$$

Eq. (2.41) naturally separates into partial sums which are characterized by fixed site indices i and j at the leftmost and rightmost single-site t-operator, respectively. Thus we define

$$\sum_{ij} \tau^{ij} = T \ , \tag{2.42}$$

with τ^{ij} the scattering path operator, first introduced by Györffy and Stott [41]. τ^{ij} comprises all possible scattering events between two sites i and j and it has to fulfill the relation

$$\tau^{ij} = t^i \delta_{ij} + \sum_{k \neq j} t^i \, G_0^{ik} \tau^{kj} \ . \tag{2.43}$$

Inserting τ into itself, this self-consistency condition in turn leads to the scattering path expansion

$$\tau^{ij} = t^i \delta_{ij} + t^i \, G_0^{ij} \, t^j + \sum_{k \neq i} \sum_{j \neq k} t^i \, G_0^{ik} \, t^k G_0^{kj} \, t^j + \dots \tag{2.44}$$

that can be used for real-space cluster calculations. In Eq. (2.44) multiple scattering events consisting of the scattering on a single scatterer n (first term), scattering between two (second term), three (third term) and more scatterers can be seen. However, Eq. (2.43) can be solved exactly obtaining the τ-matrix by matrix inversion of the KKR-matrix M:

$$\tau^{ij}(E) = \left[\underline{t}(E)^{-1} - \underline{G}_0(E)\right]_{ij}^{-1} = [\underline{M}(E)]_{ij}^{-1} \; . \tag{2.45}$$

In the case of ordered infinite systems, Eq. (2.45) can be solved exactly using the Brillouin zone integration:

$$\tau^{ij}(E) = \frac{1}{V_{\text{BZ}}} \int_{V_{\text{BZ}}} \mathrm{d}^3k \; e^{i\vec{k}(\vec{R}_i - \vec{R}_j)} \left[\underline{t}(E)^{-1} - \underline{G}_0(\vec{k}, E)\right]_{ij}^{-1} , \tag{2.46}$$

where V_{BZ} is the volume of the first Brillouin zone and $\vec{R}_{i(j)}$ denotes the lattice vector specifying the position of the unit cell $i(j)$. The $\underline{G}_0(\vec{k}, E)$ is the Fourier-transformed matrix corresponding to the free-electron Green's function $\underline{G}_0(E)$, the so-called $\vec{k}$-dependent structure constants matrix. At the zeros of the KKR-matrix determinant, the τ-matrix is singular, corresponding to a pole in the Green's function. These poles occur, according to Eq. (2.26), at the same energies as the eigenvalues of the Hamiltonian. Hence, finding the zeros of the KKR-matrix determinant gives the energy levels or the band structure of the system. To perform the Brillouin zone integration in Eq. (2.46), the Brillouin zone integration can be restricted to the irreducible part of Brillouin zone using group theory [42].

The structure constant matrix G_0 is independent on the potential within the atomic sphere and dependent on the arrangement of the potentials within the lattice. On the other hand, the single-site t-matrix is independent of the arrangement of the scattering sites and dependent on the properties of the scattering potential only. Therefore the multiple scattering problem can be divided into two independent parts. How to solve the single-site scattering on the potentials within the unit cell will be shown in the following section. The structure constants independent on the scattering potential are defined for the free electrons.

2.2.3 Solution of the single-site problem – Fully-relativistic treatment

As already mentioned above to calculate the scattering path operator τ one has to solve the single-site scattering problem. The potential V_i at site i is assumed to be spherical symmetric within the so-called Wigner-Seitz radius and constant outside

$$V_i(\vec{r}_i) = \begin{cases} V_i(r_i) & \text{for } |\vec{r}_i| < r_i^{\text{ws}} \\ 0 & \text{otherwise} \end{cases} , \qquad (2.47)$$

which makes the angular momentum representation most suitable for the treatment of this problem. As in atomic-like calculations one can separate the solutions into radial and angular components, by using spherical coordinates. In the non-relativistic case the angular dependent part of the solution is then given by spherical harmonics $Y_{lm_l}(\theta, \phi)$. The additional introduction of electron spin renders the product between spherical harmonics and the two-component Pauli spinors $Y_{lm_l}(\theta, \phi)\chi_{m_s}$ to be a suitable basis for the description of spin-dependent scattering processes. In order to account for spin-orbit coupling in addition a more complex basis is used, which will be introduced below.

2.2.3.1 The Radial Dirac Equation

Going over to a fully-relativistic treatment, Eq. (2.24) becomes in the operator notation

$$\hat{H}_{\text{D}} \,|\, \Psi \,\rangle = E \,|\, \Psi \,\rangle \;, \qquad (2.48)$$

where the Dirac Hamiltonian for a spin independent potential can be written as

$$\hat{H}_{\text{D}} = i\gamma_5 \sigma_r c \left(\frac{\partial}{\partial r} + \frac{1}{r}\left(1 - \beta\hat{K}\right) \right) + V_{\text{eff}} + (\beta - 1)\frac{c^2}{2} \;. \qquad (2.49)$$

Here, spherical coordinates and atomic units are used and the electron rest energy $\frac{1}{2}c^2$ has been subtracted. The matrices γ_5 and σ_r are given by

$$\gamma_5 = \begin{pmatrix} 0 & -I_2 \\ -I_2 & 0 \end{pmatrix} \qquad \sigma_r = \frac{1}{r}\vec{r} \cdot \vec{\sigma} \;. \qquad (2.50)$$

The spin-orbit operator $\hat{K}$ accounts for the coupling of spin and orbital angular moments and is defined by

$$\hat{K} = \vec{\sigma} \cdot \hat{\vec{L}} + 1 \ , \tag{2.51}$$

with $\hat{\vec{L}} = \hat{\vec{r}} \times \hat{\vec{p}}$ being the orbital angular momentum operator. In general, the solutions of Eq. (2.48) are bispinors in four-dimensional Dirac space, which can be separated as in the non-relativistic case, into a product of radial functions $g_\Lambda(\vec{r}, E)$ and $f_\Lambda(\vec{r}, E)$ and the so-called spin-angular functions $\chi_\Lambda(\hat{r})$

$$\Psi_\Lambda(\vec{r}) = \begin{pmatrix} g_\Lambda(r)\,\chi_\Lambda(\hat{r}) \\ if_\Lambda(r)\,\chi_{-\Lambda}(\hat{r}) \end{pmatrix} \ . \tag{2.52}$$

The indices Λ and $-\Lambda$ stand for the combined spin-orbit and magnetic quantum numbers (κ, μ) and $(-\kappa, \mu)$, respectively. The spin-angular functions $\chi_\Lambda(\hat{r})$ arise from a unitary Clebsch-Gordan transformation of the non-relativistic product functions $Y_{lm_l}(\theta, \phi)\chi_{m_s}$

$$\chi_\Lambda(\hat{r}) = \sum_{m_s = -1/2}^{+1/2} C\left(l\,\frac{1}{2}\,j; \mu - m_s, m_s\right) Y_{l,\mu - m_s}(\hat{r})\,\chi_{m_s} \ . \tag{2.53}$$

Due to the spin-orbit coupling the orbital angular momentum is not a constant of motion anymore, but only the total angular momentum operator $\hat{\vec{J}}$ still commutes with the Hamiltonian. Therefore, the spin-angular functions $\chi_\Lambda(\hat{r})$ are simultaneous eigenfunctions of the operators $\hat{\vec{J}}^2$, $\hat{J}_z$ and $\hat{K}$ with $\hat{\vec{J}} = \hat{\vec{L}} + \frac{1}{2}\vec{\sigma}$. These operators represent all constants of motion in the nonmagnetic case and their corresponding eigenvalues $j(j+1)$, κ and μ are connected by the following relations:

$$\kappa = \begin{cases} -l - 1 & \text{for } j = l + 1/2 \\ +l & \text{for } j = l - 1/2 \end{cases} \tag{2.54}$$

$$j = |\kappa| - 1/2 \tag{2.55}$$

$$-j \leq \mu \leq +j \tag{2.56}$$

$$l = \begin{cases} -\kappa - 1 & \text{for } \kappa < 0 \\ \kappa & \text{for } \kappa \geq 0 \end{cases} \tag{2.57}$$

$$\bar{l} = l - S_\kappa \ . \tag{2.58}$$

Here $S_\kappa = \kappa/|\kappa|$ is the sign of κ and $\bar{l}$ is the orbital angular momentum quantum number belonging to $\chi_{-\Lambda}$.

The presence of a magnetic field B_{eff} will break the symmetry in spin space so that only the z-component of the total angular momentum $\hat{j}_z$ will stay a constant of motion. In the case of a spin-dependent potential the Dirac Hamiltonian of Eq. (2.49) gets an additional term $\beta\sigma_z B_{\text{eff}}$.

The solutions to the Dirac equation (2.48) $\Psi_\nu(\vec{r})$ will then be superpositions of four-component spinors $\Phi_\Lambda(\vec{r})$ having different spin-angular character

$$\Psi_\nu(\vec{r}) = \sum_\Lambda \Phi_{\Lambda\nu}(\vec{r}) = \sum_\Lambda \begin{pmatrix} g_{\Lambda\nu}(r)\,\chi_\Lambda(\hat{r}) \\ if_{\Lambda\nu}(r)\,\chi_{-\Lambda}(\hat{r}) \end{pmatrix} , \qquad (2.59)$$

where ν is indexing the linearly independent solutions.

By inserting the formal solution $\Psi_\nu(\vec{r})$ into the radial Dirac Equation (2.48), renaming Λ to Λ' and carrying out a projection onto the basis $|\chi_\Lambda\rangle$ one finally arrives at the following infinite set of coupled radial differential equations

$$\frac{\partial}{\partial r}P_{\Lambda\nu}(r) = -\frac{\kappa}{r}P_{\Lambda\nu}(r) + \left[\frac{E - V_{\text{eff}}(r)}{c^2} + 1\right]Q_{\Lambda\nu}(r)$$

$$+\frac{B_{\text{eff}}(r)}{c^2}\sum_{\Lambda'}\langle\chi_{-\Lambda}\,|\,\sigma_z\,|\,\chi_{-\Lambda'}\rangle Q_{\Lambda'\nu}(r) \qquad (2.60)$$

$$\frac{\partial}{\partial r}Q_{\Lambda\nu}(r) = \frac{\kappa}{r}Q_{\Lambda\nu}(r) - [E - V_{\text{eff}}(r)]\,P_{\Lambda\nu}(r)$$

$$+B_{\text{eff}}(r)\sum_{\Lambda'}\langle\chi_\Lambda\,|\,\sigma_z\,|\,\chi_{\Lambda'}\rangle P_{\Lambda'\nu}(r) , \qquad (2.61)$$

using the notation $P_{\Lambda\nu}(r) = rg_{\Lambda\nu}(r)$ and $Q_{\Lambda\nu}(r) = crf_{\Lambda\nu}(r)$. The spin-angular matrix elements of σ_z are given by:

$$\langle\chi_\Lambda\,|\,\sigma_z\,|\,\chi_{\Lambda'}\rangle = \delta_{\mu\mu'}\begin{cases} -\dfrac{\mu}{(\kappa+1/2)} & \text{for } \kappa = \kappa' \\ -\sqrt{1 - (\frac{\mu}{\kappa+1/2})^2} & \text{for } \kappa = -\kappa' - 1 \\ 0 & \text{otherwise.} \end{cases} \quad (2.62)$$

The selection rules given in Eq. (2.62) allow only for a coupling between partial waves of the same magnetic quantum number, i.e. $\Delta\mu = 0$ and wave functions with $\Delta l = 0, \pm2, \pm4$ etc., i.e. wave functions with the same parity. Fortunately, it turns out that all couplings with $\Delta l \neq 0$ are negligibly small as their contribution is of the order of $1/c^2$. Thus, for the calculations to be presented below only the case $\Delta l = 0$ was taken into account, restricting the number of coupling terms in Eqs. (2.60) and (2.61) to at most two: $\Lambda_1 = (\kappa, \mu)$ and $\Lambda_2 = (-\kappa-1, \mu)$. Truncating the angular momentum expansion for the solutions in Eq. (2.59) at l_{max} one gets a set of $2(l_{max} + 1)^2$ linearly independent solutions $\Psi_{\Lambda\nu}(\vec{r})$ for which the radial Eqs. (2.60) and (2.61) have to be solved numerically.

2.2.3.2 Solutions of the Radial Dirac Equation

For free electrons there are two linearly independent solutions to the radial Dirac equation that represent the relativistic analogues to the well known spherical Bessel and Hankel functions and are given by [43]:

$$
j_\Lambda(\vec{r}, E) = \sqrt{1 + \frac{E}{c^2}} \left(\begin{array}{c} j_l(pr)\chi_\Lambda(\hat{r}) \\ \frac{icpS_\kappa}{E+c^2} j_{\bar{l}}(pr)\chi_{-\Lambda}(\hat{r}) \end{array} \right) \tag{2.63}
$$

$$
h_\Lambda^+(\vec{r}, E) = \sqrt{1 + \frac{E}{c^2}} \left(\begin{array}{c} h_l^+(pr)\chi_\Lambda(\hat{r}) \\ \frac{icpS_\kappa}{E+c^2} h_{\bar{l}}^+(pr)\chi_{-\Lambda}(\hat{r}) \end{array} \right) . \tag{2.64}
$$

As in the non-relativistic treatment, the relativistic Bessel functions j_Λ are regular at the origin, while the outgoing Hankel functions h_Λ^+ are irregular at the origin.

For a non-zero scattering potential the regular solution R_Λ can then be written as

$$
R_\Lambda(\vec{r}, E) = j_\Lambda(\vec{r}, E) + \int G_0(\vec{r}, \vec{r}', E)t(\vec{r}', \vec{r}'', E)j_\Lambda(\vec{r}'', E)\mathrm{d}^3r'' \tag{2.65}
$$

where the Lippmann–Schwinger equation (2.33) was used. Expanding the free electron Green's function G_0 in the basis of the eigenfunctions gives

$$
\begin{aligned}
G_0(\vec{r}, \vec{r}', E) = -ip\sum_\Lambda \big[&j_\Lambda(\vec{r}, E)h_\Lambda^{+\times}(\vec{r}', E)\Theta(r' - r) \\
&+h_\Lambda^+(\vec{r}, E)j_\Lambda^\times(\vec{r}', E)\Theta(r - r') \big] ,
\end{aligned} \tag{2.66}
$$

where the functions $h_\Lambda^{+\times}$ and $j_\Lambda^\times$ denote the left hand side or adjoint solutions of Eq. (2.48) for $V = 0$. Inserting expression (2.66) into

Eq. (2.65) results in the asymptotic form for R_Λ

$$R_\Lambda(\vec{r}, E) = j_\Lambda(\vec{r}, E) + ip \sum_{\Lambda'} h_{\Lambda'}^+(\vec{r}, E) t_{\Lambda'\Lambda}(E) \ , \qquad (2.67)$$

i.e. for $r > r^{\mathrm{ws}}$ a solution must approach a superposition of a partial wave j_Λ with pure spin-angular character Λ and an outgoing spherical wave, whose spin-angular composition is given by the single-site t-matrix $t_{\Lambda'\Lambda}$. This $t_{\Lambda'\Lambda}$ matrix describes the scattering of states with character Λ into the scattering channels Λ'.

Solving Eqs. (2.60) and (2.61) by numerical outward integration will yield the linearly independent regular solutions Ψ_ν of mixed spin-angular character, from which any other regular solution R_Λ can be constructed by

$$R_\Lambda(\vec{r}, E) = \sum_\nu A_{\Lambda\nu} \Psi_\nu(\vec{r}, E) \ . \qquad (2.68)$$

The coefficients $A_{\Lambda\nu}$ can then be determined from the asymptotic behavior of the R_Λ's and Ψ_ν's.

Ebert and Györffy [44] showed, that the relativistic transition matrix can be expressed as

$$t_{\Lambda\Lambda'} = \frac{i}{2p} \left[(a - b) b^{-1} \right]_{\Lambda\Lambda'} \ , \qquad (2.69)$$

where the matrices $a_{\Lambda\Lambda'}$ and $b_{\Lambda\Lambda'}$ are given by

$$a_{\Lambda\Lambda'}(E) = -ipr^2 \left[h_\Lambda^-(\vec{r}, E), \Phi_{\Lambda\nu}(\vec{r}, E) \right]_r \qquad (2.70)$$

$$b_{\Lambda\Lambda'}(E) = ipr^2 \left[h_\Lambda^+(\vec{r}, E), \Phi_{\Lambda\nu}(\vec{r}, E) \right]_r \ . \qquad (2.71)$$

Here $p = \sqrt{E(1 + E/c^2)}$ is the relativistic momentum [43] and the term in parenthesis denotes the relativistic form of the Wronskian expression [44]:

$$\left[h_\Lambda^+, \Phi_{\Lambda\Lambda'} \right]_r = h_l^+ c f_{\Lambda\Lambda'} - \frac{p}{1 + E/c^2} S_\kappa h_{\bar{l}}^+ g_{\Lambda\Lambda'} \ . \qquad (2.72)$$

Using these results for the single-site t-matrix, it is now possible to write down the Green's function for the single-site scattering problem by rewriting Eq. (2.36)

$$G^i = G_0 + G_0 t^i G_0 \ . \qquad (2.73)$$

Inserting Eq. (2.66) in Eq. (2.73) one finally obtains the single-site Green's function for site i in the form

$$
\begin{aligned}
G^i(\vec{r}, \vec{r}', E) &= \sum_{\Lambda, \Lambda'} Z^i_\Lambda(\vec{r}, E) t^i_{\Lambda\Lambda'}(E) Z^{i\times}_{\Lambda'}(\vec{r}', E) \\
&\quad - \sum_\Lambda Z^i_\Lambda(\vec{r}, E) J^{i\times}_\Lambda(\vec{r}', E) \Theta(r' - r) \\
&\quad - \sum_\Lambda J^{i\times}_\Lambda(\vec{r}, E) Z^i_\Lambda(\vec{r}', E) \Theta(r - r') \, , \quad (2.74)
\end{aligned}
$$

where the functions Z_Λ are given by

$$
\begin{aligned}
Z_\Lambda(\vec{r}, E) &= \sum_{\Lambda'} R_\Lambda(\vec{r}, E) t^{-1}_{\Lambda'\Lambda} \quad\quad\quad\quad\quad (2.75) \\
&= \sum_{\Lambda'} j_{\Lambda'}(\vec{r}, E) t^{-1}_{\Lambda'\Lambda}(E) - i p h^+_\Lambda(\vec{r}, E) \, . \quad (2.76)
\end{aligned}
$$

Here, the functions J_Λ are the irregular solutions to the Dirac equation obtained by inward integration of the radial equations, initialized by the boundary condition

$$
J_\Lambda(\vec{r}, E) = j_\Lambda(\vec{r}, E) \quad \text{for } r \geq r^{\mathrm{ws}} \, . \quad (2.77)
$$

Furthermore, with all single-site t-matrices one can set up the super matrix $\underline{t}$ and calculate $\underline{\tau}$ for the whole system by using Eq. (2.45). The Green's function for the whole system is then given by

$$
G = G_0 + \sum_{i,j} G_0 \tau^{ij} G_0 \, , \quad (2.78)
$$

where the Dyson equation (2.36) and the definition of the scattering path operator Eq. (2.42) have been used. The final relativistic expression for G can then be written as follows:

$$
\begin{aligned}
G(\vec{r}_i + \vec{R}_i, \vec{r}'_j + \vec{R}_j, E) &= \sum_{\Lambda, \Lambda'} Z^i_\Lambda(\vec{r}_i, E) \tau^{ij}_{\Lambda\Lambda'}(E) Z^{j\times}_{\Lambda'}(\vec{r}'_j, E) \\
&\quad - \delta_{ij} \sum_\Lambda \Big[Z^i_\Lambda(\vec{r}_i, E) J^{i\times}_\Lambda(\vec{r}'_i, E) \Theta(r'_i - r_i) \\
&\quad + J^i_\Lambda(\vec{r}_i, E) Z^{i\times}_\Lambda(\vec{r}'_i) \Theta(r_i - r'_i) \Big] \, . (2.79)
\end{aligned}
$$

As already mentioned in Sec. 2.2.1, with G one can then calculate all single-electron properties as for example the particle spin density which is needed for the self-consistent DFT cycle

$$
n(\vec{r}) = -\frac{1}{\pi} \operatorname{Im} \int^{E_{\mathrm{F}}} \mathrm{d}E \operatorname{Tr} G(\vec{r}, \vec{r}, E) \, . \quad (2.80)
$$

The expectation values of all other observables can be obtained using Eq. (2.30).

2.2.4 Treatment of disordered alloys

To be able to describe disordered alloys, the theoretical methods presented so far have to be reconsidered, because of their restriction to ordered stoichiometric systems. In practise growing a layered system on top of a substrate often leads to interdiffusion at the interface. The interdiffusion between layers could be modeled by random substitutional alloys at the neighboring layers. Such a disordered material is assumed to possess structural order. However, due to the random distribution of its constituent atoms, such a material is not translational invariant and Bloch's theorem cannot be applied directly.

A viable idea is to evade this problem and to replace the disordered system by an ordered one consisting of "effective atoms". Within the Virtual Crystal Approximation (VCA), for example, an average potential is used to calculate the effective scattering [45]. By averaging the t-matrix of atomic species forming the alloy, one obtains the Average t-matrix Approximation (ATA) [46].

A much more reliable description can be achieved on the basis of the Coherent Potential Approximation (CPA) alloy theory. CPA theory is considered to be the best theory among the so-called single-site alloy theories that assume complete random disorder and ignore short-range order [47]. Combining the CPA with multiple scattering theory leads to the KKR-CPA scheme, which is nowadays intensively applied for quantitative investigations of the electronic structure and properties of disordered alloys [47, 48]. Within the CPA configurationally averaged properties of a disordered alloy are represented by the hypothetical ordered CPA-medium, which in turn may be described by a corresponding scattering path operator τ_{CPA}^{ii}. The corresponding single-site t-matrix t_{CPA} and multiple scattering path operator τ_{CPA}^{ii} are determined by the so called CPA-condition:

$$x_{\mathrm{A}}\, \tau_{\mathrm{A}}^{ii} + x_{\mathrm{B}}\, \tau_{\mathrm{B}}^{ii} = \tau_{\mathrm{CPA}}^{ii} \ . \tag{2.81}$$

Hence the scattering path operator of the CPA-medium is concentration weighted (x_{α}, $\alpha =$ A,B) sum of the component-projected scattering path operators $\underline{\tau}_\alpha^{ii}$, $\alpha = \text{A}, \text{B}$. The above equation represents the requirement that embedding an atom (of type A or B) into the CPA medium substitutionally should not cause additional scattering. The scattering properties of an A atom embedded in the CPA medium are represented by the component-projected scattering path operator $\underline{\tau}_\text{A}^{ii}$

$$\underline{\tau}_\text{A}^{ii} = \underline{\tau}_\text{CPA}^{ii} \left[1 + \left(\underline{t}_\text{A}{}^{-1} - \underline{t}_\text{CPA}^{-1}\right) \underline{\tau}_\text{CPA}^{ii}\right]^{-1} , \qquad (2.82)$$

where $\underline{t}_\text{A}$ and $\underline{t}_\text{CPA}$ are the single-site matrices of the A component and of the CPA medium. Because $\underline{\tau}_\text{CPA}^{ii}$ is determined by Eq. (2.45), with $\underline{t}$ replaced by $\underline{t}_\text{CPA}$, these coupled sets of equations for $\underline{\tau}_\text{CPA}^{ii}$ and $\underline{t}_\text{CPA}$ have to be solved iteratively.

2.3 Tight-binding KKR (TB-KKR) method

For fast calculations of large layered systems the standard KKR method suffers from the fact that the computational time increases cubically with the system size (N^3). In this respect a major breakthrough has been achieved by the concept of screening [49], by which the KKR method can be transformed into a tight-binding form with short-range interactions between the atoms. The resulting TB-KKR-matrix has a band form, unlike the standard "full" and infinite KKR-matrix. Thanks to the band form of the TB-KKR-matrix, the computational time of the TB-KKR method increases only linearly with the system size (N). This fact makes it possible to proceed calculations of large systems within a reasonable time.

One way to develop the tight-binding form of the KKR method goes back to the work of Andersen *et al.* [49], who showed that by generalizing the screening concept of the tight-binding LMTO method to energy-dependent screening parameters [50] and by optimizing the screening parameters exponentially decaying "screened" structure constants can be obtained. Based on this idea Szunyogh *et al.* [51] formulated and successfully applied a screened KKR method for surfaces [52]. Unfortunately, the screened structure constants obtained according to [51] do not decay sufficiently fast, thus limit the efficiency of the method.

Another method to obtain the tight-binding form of the KKR method, which is physically and mathematically simpler and more transparent, is based on Green's function formalism and the concept of a repulsive reference system. This concept was introduced by a collaboration of the Jülich and the Vienna/Budapest group [53]. For a suitably chosen reference system, for instance, consisting of an infinite array of repulsive spherical wells, the resulting structure constants decay exponentially for energies sufficiently below the top of the well. The demonstration and test of this appropriate reference system is discussed by Wildberger [54]. Several authors implemented the fully-relativistic formalism into the TB-KKR code, allowing calculations of spin-orbit induced phenomena in layered condensed materials. For the sake of simplicity, however, non-relativistic notation will be used in this section.

2.3.1   Transformation to the tight-binding form

The tight-binding KKR method is basically a reformulation of multiple scattering theory in terms of exponentially decaying structure constants, which can be obtained by a screening transformation. This tight-binding KKR method has in principle the same accuracy like the standard KKR method. Anyway, it is important to point out that including a limited number of atoms in the tight-binding cluster for calculation of the structure constants leads to a cut off in space. Therefore a sufficient number of atoms in the tight-binding cluster has to be established. On the other hand, the advantage of the screened structure constants is that they are easy to use and have no singularities. The biggest advantage of the new method is that it allows N-scaling calculations. It is particularly well suited for nanostructures like surfaces, multilayers, impurities in the bulk or on surfaces etc.

The main idea of the chosen approach is based on the freedom in choosing the reference system [36, 55]. The KKR method in the original form, which was particularly used for calculation of the band structure of infinite solid state materials, uses the free space as a natural reference system. It will be shown that the standard KKR method can be formulated with respect to any reference system and that a transformation of the standard structure constants of the free space into a tight-binding form is possible.

2.3.2   Transformation to an arbitrary reference system

The structural Green's function matrix $\underline{G}(E) = \left\{ G_{LL'}^{ij}(E) \right\}$ can be expressed through the Dyson equation in terms of the reference function of the free space $\underline{G}^0(E) = \left\{ G_{LL'}^{0\,ij}(E) \right\}$:

$$\underline{G} = \underline{G}^0 + \underline{G}^0 \underline{t}\, \underline{G}^0 + \underline{G}^0 \underline{t}\, \underline{G}^0 \underline{t}\, \underline{G}^0 + \dots = \underline{G}^0 \cdot (\underline{I} - \underline{t}\, \underline{G}^0)^{-1} \ . \quad (2.83)$$

$\underline{G}^0(E)$ are also called the free structure constants. $\underline{t}(E) = \left\{ t_{LL'}^i(E)\delta_{ij} \right\}$ is the "atomic" t-matrix and $\underline{I}$ stands for the unitary matrix in the angular momentum representation. For the sake of simplicity, the energy dependence of the matrices in Eq. (2.83) has been suppressed.

Using the scattering path operator $\underline{\tau}(E) = \left\{ \tau_{LL'}^{ij}(E) \right\}$

$$\underline{\tau} = \left[\underline{t}^{-1} - \underline{G}^0 \right]^{-1} \ , \quad (2.84)$$

Eq. (2.83) can be written as follows:

$$\underline{G} = \underline{G}^0 + \underline{G}^0 \underline{\tau}\, \underline{G}^0 \ . \quad (2.85)$$

Eqs. (2.83)–(2.85) build the basis of the standard KKR method. As already mentioned, these Eqs. can be reformulated with respect to a new reference system, characterized in the following by the index r. In this system r, new arbitrary scattering potentials V^r will be localized at the scattering centers. With $\underline{t}^r(E)$-matrix being the t-matrix and $\underline{G}^r(E)$ being the structural Green's function matrix of the reference system r one obtains analogously to the Eq. (2.83)

$$\underline{G}^r = \underline{G}^0 + \underline{G}^0 \underline{t}^r\, \underline{G}^0 + \underline{G}^0 \underline{t}^r\, \underline{G}^0 \underline{t}^r\, \underline{G}^0 + \dots = \underline{G}^0 \cdot (\underline{I} - \underline{t}^r\, \underline{G}^0)^{-1} \ . \quad (2.86)$$

Up to now no proposition about the kind of the reference system has been done. Defining the difference of the t-matrices

$$\Delta \underline{t} = \underline{t} - \underline{t}^r \quad (2.87)$$

the Green's function of the real system $\underline{G}$ can be related to the Green's function of the reference system $\underline{G}^r$:

$$\underline{G} = \underline{G}^r + \underline{G}^r \Delta \underline{t}\, \underline{G}^r + \underline{G}^r \Delta \underline{t}\, \underline{G}^r \Delta \underline{t}\, \underline{G}^r + \dots = \underline{G}^r \cdot (\underline{I} - \Delta \underline{t}\, \underline{G}^r)^{-1} \ . \quad (2.88)$$

Analogously to the Eq. (2.84), the scattering path operator will be described as

$$\underline{\underline{\tau}}_\Delta = \left[(\Delta \underline{t})^{-1} - \underline{G}^r \right]^{-1} \ . \quad (2.89)$$

Then the Green's function $\underline{G}$ can be again reformulated in respect to the Green's function of the reference system $\underline{G}^r$:

$$\underline{G} = \underline{G}^r + \underline{G}^r \underline{t}_\Delta \underline{G}^r \ . \tag{2.90}$$

As soon as the $\underline{t}^r$ and $\underline{G}^r$ are known, the relations (2.88)–(2.90) give a set of equations equivalent to Eqs. (2.83)–(2.85).

This way it was shown that the Dyson equation can be formulated with respect to any arbitrary reference system.

2.3.3 Tight-binding KKR method for slab geometry

In the previous chapter the transformation of the KKR method to a tight-binding form was shown. Choosing a reference system with constant repulsive muffin-tin potentials at the lattice positions, structure constants exponentially decaying in real space can be achieved. Due to the finite range of the screened structure constants, the computational effort scales then linearly with the size of the system, not cubically like in the standard KKR method.

As the screened structure constants decay very fast, they can be calculated in real space using a cluster of atoms. Because of the two-dimensional periodicity in the plane of the layered system, screened structure constants will be Fourier transformed in two dimensions:

$$G_{LL'}^{r\,ij}(\vec{k}_\parallel, E) = \sum_{\nu'} \exp\left[-i\vec{k}_\parallel(\vec{\chi}_\nu - \vec{\chi}_{\nu'})\right] G_{LL'}^{r\,ij,\,\nu-\nu'}(E) \ . \tag{2.91}$$

To be able to formulate the Dyson equation for a layered system a suitable indexing of atoms will be adopted. All monolayers parallel to the surface will obtain a layer index i, all atoms inside a monolayer will be characterized by an index ν. An arbitrary lattice vector $\vec{R}^n$ can be decomposed to $\vec{R}^n = \vec{R}_i + \vec{\chi}_\nu$. $\vec{R}_i$ describes the position of the origin within the i-th monolayer and $\vec{\chi}_\nu$ is an arbitrary vector of the two-dimensional lattice that is identical for each monolayer. In Fig. 2.1 the reference cluster in real space with a layer adjusted notation is shown.

Because of the exponential decay of the screened structure constants the summation in the Eq. (2.91) runs over a finite number of two-dimensional lattice vectors $\vec{\chi}_{\nu'}$ only. The three-dimensional structure constants $G_{LL'}^{r\,ij,\,\nu-\nu'}(E)$ are stored and the $\vec{k}_\parallel$-dependent screened structure constants always calculated when needed.

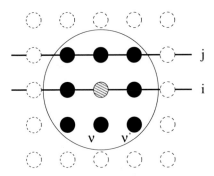

Fig. 2.1. Schematic description of a cluster in real space. The different layers have an index i, different atoms within one layer are labeled by the index ν.

The Dyson equation has after two-dimensional Fourier transformation the form as for a linear chain:

$$
\begin{aligned}
G^{ij}_{LL'}(\vec{k}_{\|}, E) &= G^{r\,ij}_{LL'}(\vec{k}_{\|}, E) \\
&+ \sum_{nL''} G^{r\,in}_{LL''}(\vec{k}_{\|}, E)\,(t^n_{L''}(E) - t^{r\,n}_{L''}(E))\,G^{nj}_{L''L'}(\vec{k}_{\|}, E) \quad (2.92)
\end{aligned}
$$

The finite layered system, for which the Dyson-equation (2.92) is solved, is sketched in the Fig. 2.2. The thick solid lines present the material layers and the dashed lines demonstrate the "vacuum layers" with potential zero.

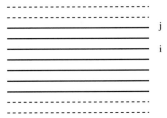

Fig. 2.2. Schematic description of a slab. The thick solid lines present the material layers and the dashed lines demonstrate the "vacuum layers" with potential zero.

The Fourier-transformed Green's function of the reference system $G^{r\,ij}_{LL'}(\vec{k}_{\|}, E)$, as well as the $\underline{I} - (\underline{t} - \underline{t}^r)\underline{G}^r$ matrix which has to be inverted in Eq. (2.89), have a band form. The band form appears due to the restriction to the coupling with the nearest neighbors. Then only the diagonal $(j = i)$ and the neighboring upper and lower layers $(j = |i\pm n|)$ couple together. n is the number of neighboring layers included for coupling with the layer i. If one includes the coupling between nearest layers only, the resulting band form looks like a tight-binding matrix (x means a non-zero contribution, whereas 0 denotes no interaction):

$$\begin{pmatrix} x & x & 0 & 0 & 0 & 0 \\ x & x & x & 0 & 0 & 0 \\ 0 & x & x & x & 0 & 0 \\ 0 & 0 & x & x & x & 0 \\ 0 & 0 & 0 & x & x & x \\ 0 & 0 & 0 & 0 & x & x \end{pmatrix} . \tag{2.93}$$

It is important to point out that the coupling between the layers occurs through the screened structure constants $\underline{G}^{r\,ij}$ only.

In the following chapter it will be shown that this way a N-scaling procedure for layered systems can be implemented.

2.3.4 N-scaling procedure

Because the computational time in the standard KKR method with three-dimensional translational invariance increases cubically with the system size (N^3), the size of the calculated systems is limited. Using the screened KKR method leads to a final range of the structure constants and the computational time scales linearly with the number of layers. The particular advantage of the screened KKR method lies in the enormous saving of computational time. The solution of the Dyson equation (2.90) requires the inversion of the $(\underline{I} - \Delta\underline{t}\,\underline{G}^r)$ matrix, as shown in Eq. (2.88). The Green's reference function $\underline{G}^r$ has a block matrix form and therefore the solution can be achieved very fast and efficiently. Further the $\Delta\underline{t}$ matrix is diagonal in the layer indices. When calculating the diagonal elements of the Green's function $G_{LL'}^{ii}(\vec{k}_\parallel, E)$ according to the algorithm of Godfrin [56], the computational effort scales linearly with the number of considered layers N.

The central variable in the DFT is the density $n(\vec{r})$, which can be calculated in every cell i from the diagonal element G^{ii} and that is why in self-consistent calculations the computer time scales linearly with N. In contrast, the determination of the full matrix $G^{ij}_{LL'}(\vec{k}_{\parallel}, E)$ requires N^2 operations. However, this full matrix is not needed for the self-consistent solution of the Kohn-Sham equations. As follows, very big layered systems can be dealt with.

2.4 Magnetic anisotropy energy (MAE)

Ferromagnetic materials exhibit intrinsic "easy" and "hard" directions of the magnetization. The "easy" axis denotes the spontaneous magnetization direction and the system lies in the energy minimum. The "hard" axis shows the direction of magnetization with the system being in the energy maximum. This magnetic anisotropy is, both from a technological and fundamental point of view, one of the most important properties of magnetic materials. Depending on the type of application, materials with high, medium or low magnetic anisotropy will be required, for respective application as, e.g., permanent magnets, information storage media or magnetic cores in transformers, and magnetic recording heads. Hence a good understanding of the microscopic origin of the magnetic anisotropy is necessary to tailor the properties of magnetic materials. The magnetic anisotropy can be investigated by various experimental techniques (magnetometry, Brillouin light scattering etc.) or by calculating the energy of systems with various magnetization directions and finding the one with the lowest energy.

The magnetic anisotropy energy (MAE) is the energy change due to a change of the magnetization direction and determines the spontaneous direction of the magnetization. Conventionally, the MAE is split into an electronic band structure part (magnetocrystalline anisotropy energy ΔE_{SO}) that is caused by the spin-orbit coupling and a part (the shape anisotropy energy ΔE_{d}) that stems from the classical dipole-dipole interaction of the magnetic moments. van Vleck already pointed out that the physical origin of the electronic magnetocrystalline anisotropy energy and the orbital moment is the spin-orbit coupling [1]. Brooks and Fletcher applied the ideas of van Vleck to itinerant ferromagnets using a semi-empirical band structure model [57, 58]. Nowadays it is well established that the electronic contribution and the orbital moment come from the simultaneous occurrence of spin-orbit coupling and spin polarization. For that reason, calculations of the orbital moment and the electronic magnetocrystalline anisotropy energy of bulk materials [59–67], transition metal alloys [68–70], transition metal film monolayers and multilayers [64, 71–77] and semiconductor/ferromagnet heterogeneous layered systems [78–81] from first principles became possible only after the development of spin-polarized relativistic band structure methods. Whereas in most of the mentioned studies the spin-orbit coupling was included as perturbation, the strength of the following work lies in the fact that the band structure has been calculated in a fully-relativistic way (see Sec. 2.2.3).

2.4.1 Magnetocrystalline anisotropy energy

The electronic MAE is determined as the difference in total energy of the system for two different orientations of the magnetization $\vec{M}$. For example for the difference in total energy of an out-of- and in-plane orientation of $\vec{M}$ one has

$$\Delta E_{\mathrm{SO}} = E(\vec{M} \parallel \hat{c}) - E(\vec{M} \perp \hat{c}) , \qquad (2.94)$$

with the vector $\hat{c}$ perpendicular to the layers. Because the calculation of the MAE via the total energy requires a self-consistent calculation for both orientations of $\vec{M}$ corresponding studies are computationally rather demanding. One therefore often uses the so-called Force theorem to determine the MAE [82, 83]. In this case the electronic potential has to be calculated self-consistently only for one orientation of $\vec{M}$. Keeping the potential fixed, the MAE is then determined as the difference of the band or single particle energies obtained from calculations for the two different orientations of $\vec{M}$. We have checked the accuracy of the simplified Force theorem–based approach and found in general deviations less than 10% from the results based on a full total energy calculation.

In the Force theorem method the number of $\vec{k}$ points within the full Brillouin zone has to be very large in order to obtain a stable (converged) value of the anisotropy energy. As reported in literature the Force theorem MAE value for transition metal thin films is stable when sampling is done for, at least, 10^4 $\vec{k}$ points within the full three-dimensional Brillouin zone [73, 64, 71, 72, 74, 75]. For these films the anisotropy energy has typically a value of about 10^{-4}–10^{-5} Ry/atom. This is the same order of magnitude as for the layered systems treated in the following chapters. The band energies and the electronic anisotropy energy reach stable values for about 10^3 $\vec{k}$ points in the full two-dimensional Brillouin zone using the two-dimensional tight-binding KKR method (see Sec. 2.3).

2.4.2   Shape anisotropy

To determine the shape anisotropy contribution to the MAE in a layered system one can evaluate the corresponding Madelung sums directly in real space. In the following the shape anisotropy contributes only when comparing the out-of- and in-plane orientation of $\vec{M}$. When comparing two orientations in the plane of the systems considered here that are perpendicular to one another, the shape anisotropy does not contribute due to the square two-dimensional symmetry of these layered systems. If all the magnetic moments of the atoms are parallel to the direction $\hat{n}$, then the magnetic dipole-dipole interaction energy for this direction is given, in atomic Rydberg units, by

$$E_d(\hat{n}) = \sum_{q,q'} \frac{m_q m_{q'}}{c^2} M_{qq'} \tag{2.95}$$

$$M_{qq'} = \sum_{\vec{R}} \frac{1}{|\vec{R} + \vec{q} - \vec{q}'|^3}\left(1 - 3\frac{[(\vec{R} + \vec{q} - \vec{q}') \cdot \hat{n}]^2}{|\vec{R} + \vec{q} - \vec{q}'|^2}\right), \tag{2.96}$$

where c is the speed of light (274.072 in Rydberg units), $\vec{q}$ and $\vec{q}'$ denote the atom positions in one unit cell, m_q is the total magnetic moment in an atomic sphere around site $\vec{q}$ and $\vec{R}$ is the translation lattice vector. $|\vec{R} + \vec{q} - \vec{q}'|$ represents the distance between the atom $\vec{q}$ and all other atoms $\vec{q}'$ in the system. The sum runs over all the lattice sites at $\vec{R}$ except over that for which the denominator in Eq. (2.96) is zero. The dipole-dipole contribution to the MAE, Eq. (2.94), is the difference between $E_d(\vec{M} \parallel \hat{c})$ and $E_d(\vec{M} \perp \hat{c})$.

2.4.3   Bruno's and van der Laan's formulas

In the past several authors pointed out that one can connect the MAE with the anisotropy of other spin-orbit induced properties. The most prominent example for this is the interrelation of the MAE and the anisotropy of the orbital angular momentum, that was considered by Bruno [84] on the basis of second order perturbation theory. While Bruno assumed a fully occupied majority spin band, this restriction was dropped in the later work of van der Laan [85], who derived the relation

$$\Delta E_{\rm SO} = -\frac{\xi}{4}\hat{n}_{\parallel}(L_{\parallel}^{\downarrow} - L_{\parallel}^{\uparrow}) + \frac{\xi}{4}\hat{n}_{\perp}(L_{\perp}^{\downarrow} - L_{\perp}^{\uparrow}) \,. \qquad (2.97)$$

Here ξ is the spin-orbit coupling parameter and $L_{\hat{n}}^{\uparrow(\downarrow)}$ is the orbital moment of the majority (minority) spin band with the subscript $\hat{n}$ indicating that $L^{\uparrow(\downarrow)}$ is determined for the orientation of $\vec{M}$ along $\hat{n}$. According to the Eq. (2.94), the symbol $\parallel$ represents the magnetization direction parallel to the vector $\hat{c}$ being perpendicular to the layer plane and the symbol $\perp$ represents the in-plane magnetization direction. $\hat{n} \cdot L_{\hat{n}}^{\uparrow(\downarrow)}$ gives the component of $L_{\hat{n}}^{\uparrow(\downarrow)}$ along the magnetization $\vec{M}$. Ignoring the majority band ($\uparrow$) contribution in Eq. (2.97) and setting $L_{\hat{n}}^{\downarrow} = L_{\hat{n}}$ one ends up with the expression given by Bruno [84]:

$$\Delta E_{\rm SO} = -\frac{\xi}{4} \left(\hat{n}_{\parallel} L_{\parallel} - \hat{n}_{\perp} L_{\perp} \right) \,. \qquad (2.98)$$

For an application of the expressions of Bruno and van der Laan, we used for all atoms the corresponding bulk values of the spin-orbit coupling parameter ξ. Having access to both calculated quantities, MAE and the anisotropy of the orbital moments, one can check directly the validity of Bruno's and van der Laan's models, as will be shown in Sec. 5.4.

Chapter 3

Fe surfaces and clusters

3.1 Introduction

The possible appearance of a large magnetocrystalline surface anisotropy was pointed out already 50 years ago by Néel [86]. In the last years, thin Fe films have been investigated in several experiments [87, 88] and a magnetic anisotropy energy (MAE) several magnitudes higher than in the Fe bulk system has been detected. In order to determine the correct spontaneous magnetization direction by calculations, the importance of the dipole-dipole part to the MAE (shape anisotropy) will be demonstrated in the Sec. 3.2.4. Anyway, to understand the magnetic properties of Fe surfaces, one has to investigate the distribution of magnetic moments in the layers (magnetization profiles) first. Several calculations of the dependence of μ_{spin} on the depth below the crystal surface have been published for bcc Fe, using an *ab initio* formalism [89–95] as well as a TB model Hamiltonian [96]. Most of these investigations deal with the (001) surface; less work has been done on the (110) surface and only little attention has been devoted to the (111) surface. Not many papers include μ_{orb} in their consideration. *Ab initio* calculations of the profile of μ_{orb} were published for a perpendicularly (out-of-plane) magnetized (001) surface [91, 95, 97], a systematic study of the orbital magnetism below surfaces of transition metals was performed by Rodríguez-López *et al.* [96].

Clusters comprising few tens or hundreds of atoms form an interesting class of materials, because they form a bridge between atoms and molecules on the one hand and solids on the other hand and yet their properties cannot be described by a simple interpolation between the two extremes. Consequently, magnetic properties of transition metal clusters attracted a lot of attention – both due to fundamental reasons and due to a potential application in magnetic recording industry. As clusters contain a large portion of surface atoms, it is interesting to study the relation between the electronic and magnetic properties of atoms which are close to a cluster surface and of atoms which are close to a planar surface of a crystal. All calculations on the Fe clusters have been performed by Šipr. The results of our collaboration have been already published [98–100] and with the permission of Šipr also used in this thesis. Although the importance of surface-related effects in clusters has been universally acknowledged, no systematic study comparing clusters and crystal surfaces has been performed so far, to the best of our knowledge. One of the aims of this work is thus to focus on theoretical investigations on free iron clusters with bcc geometry and bulk interatomic distances and on comparing their properties with properties of bcc Fe crystal surfaces.

Previous work on free medium-sized Fe clusters of 10–100 atoms with a geometry taken as if cut from the bulk relied mostly on a parametrized tight-binding (TB) model Hubbard Hamiltonian [101–103]. Early *ab initio* calculations, on the other hand, were restricted to Fe clusters containing not more than 15 atoms [104–106]. More recent work relying on an *ab initio* approach have focused mainly on geometry optimization of small or medium-sized clusters [107, 108] and thus cannot be directly utilized for comparing with crystal surfaces. Generally, nearly all calculations of electronic structure of free metallic clusters were non-relativistic, meaning that they do not give access to the orbital contribution to the magnetic moment. Only recently orbital magnetic moments for free Ni clusters obtained by means of TB model calculations with spin-orbit coupling included via an intra-atomic approximation were presented [109].

The purpose of this chapter is to investigate theoretically magnetic properties of Fe surfaces and free Fe clusters. In the next section magnetic moments, DOS and the magnetic anisotropy of Fe surfaces are discussed in detail. Moreover, the dependence of μ_{spin} and μ_{orb} on the distance of atomic sites from the cluster center and the dependence of μ_{orb} on the direction of the magnetization is discussed for several

cluster sizes. This is followed by a comparison of the magnetic properties of atoms in free clusters to properties of atoms at and below
crystal surfaces. Then systematic trends of magnetic moments in clusters and at surfaces and in particular their dependence on the effective
coordination number are studied.

3.2 Fe surfaces

3.2.1 Structure and computational details

Crystal surfaces have been simulated by finite 2D slabs, with bulk interatomic distances throughout the whole system. For the (001) and
(110) surface 18 layers of Fe atoms and for the (111) surface 34 layers
of Fe atoms have been used. These investigated surfaces of the bcc Fe
crystal are depicted in Fig. 3.1, where the dashed lines denote the surface plane. In order to account for the spilling of electron charge into
the vacuum, all slabs have been surrounded by layers of empty spheres.
The electronic and magnetic structure of these slabs has been calculated
via the fully-relativistic spin-polarized tight-binding Korringa-Kohn-
Rostoker (TB-KKR) method [51, 53] described in Sec. 2.3. We relied
on spherical potentials in the atomic sphere approximation (ASA).

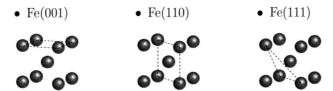

- Fe(001)   • Fe(110)   • Fe(111)

Fig. 3.1. Structure of bcc Fe surfaces. From left to right: Fe(001), Fe(110) and Fe(111)
surface. The dashed lines denote the surface plane.

3.2.2 Densities of states

In Fig. 3.2 densities of states (DOS) of Fe atoms with (001) and (111)
oriented surfaces are shown. Because of the smaller coordination number of the atoms at the surface (L=0) a sharpening of the DOS appears.

Only in the fifth layer from the surface (L=4) the bulk-like DOS can be observed for both (001) and (111) slabs. Convergence to the bulk properties in the center of the slab (L=8 in the Fe(001) system and L=18 in the Fe(111) system) denotes the sufficient size of the slabs describing a semi-infinite crystal.

In the case of the (110) surface, surface atoms have 6 nearest neighbors, whereas in the central layers the number of nearest neighbors is 8. That leads to a sharpening of the DOS of the surface atoms, which is anyway not so significant as in the (001) and (111) surfaces with only 4 nearest neighbors for the surface atoms. As shown in Fig. 3.3 in the (110) system already the second subsurface layer (L=2) shows a bulk-like DOS.

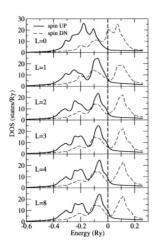 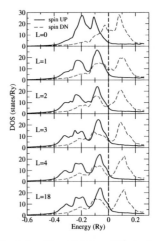

Fig. 3.2. In the left panel spin- and layer-resolved densities of states of Fe in the (001) oriented surface system are shown. Densities of states of Fe atoms in the Fe(111) surface system are shown in the right panel. The surface layer is denoted by "L=0", the subsurface layer by "L=1", the underlying layer by "L=2" and so on. The lowest picture shows the layer in the center of the slab. The solid (dashed) line represents the spin up (spin down) states.

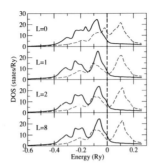

Fig. 3.3. Densities of states of Fe atoms in the Fe(110) surface system. The surface layer is denoted by "L=0", the subsurface layer by "L=1", the underlying layer by "L=2" and so on. The lowest picture shows the layer in the center of the slab. The solid (dashed) line represents the spin up (spin down) states.

3.2.3   Magnetic properties

In this section we display the results of our calculations of μ_{spin} and μ_{orb} at Fe crystal surfaces and compare them with available theoretical results obtained via different methods. Fig. 3.4 summarizes the layer-dependence of μ_{spin} for the (001), (110) and (111) crystal surfaces and compares them with full-potential linearized augmented plane wave (FP-LAPW) method calculations of Ohnishi et al. [89] and Fu and Freeman [90], linear-muffin-tin-orbital (LMTO) calculations of Eriksson et al. [97], full-potential LMTO calculations of Hjortstam et al. [91], TB-LMTO calculations of Spišák and Hafner [92], Green's function LMTO calculations of Niklasson et al. [93], linear combination of (pseudo) atomic orbitals calculations of Izquierdo et al. [94] and d-band model Hamiltonian calculations of Rodríguez-López et al. [96].

One can see that the basic trend, namely, a rather strong enhancement of μ_{spin} at crystal surfaces, is attained by all calculations. For the (110) surface this enhancement is significantly smaller than for the (001) and (111) surfaces; this is consistent with the fact that atoms at the (110) surface have higher coordination numbers than atoms at the other two surfaces. The convergence of μ_{spin} to bulk values has been attained at the center of the slabs. The increase of μ_{spin} with decreasing distance from the surface is usually not monotonous. It was suggested that these

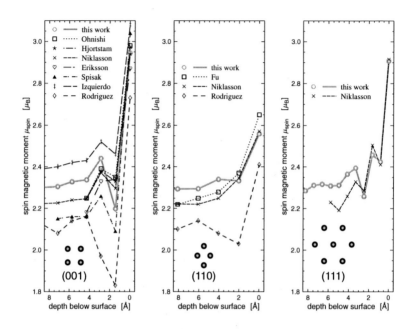

Fig. 3.4. Layer-dependence of μ_{spin} below the (001) Fe crystal surface (left panel), (110) crystal surface (middle panel), and (111) crystal surface (right panel). The surface structures are schematically depicted in the insets. Together with our results we display results of Ohnishi *et al.* [89], Fu and Freeman [90], Eriksson *et al.* [97], Hjortstam *et al.* [91], Spišák and Hafner [92], Niklasson *et al.* [93], Izquierdo *et al.* [94] and Rodríguez-López *et al.* [96], as indicated in the legend.

Friedel-like oscillations are an artifact caused by an insufficient number of layers involved in slab-type calculations [89, 90]. However, these oscillations persist even if the number of layers in which the electronic structure has been allowed to relax is as large as nine or ten (like in our work or in Ref. [93]) and also when dealing with a semi-infinite crystal geometry [95]. Likewise, these oscillations are present in ASA as well as in full-potential calculations [89, 91, 94], so approximating the shape of the potential does not seem to be significant in this respect either. Note that full-potential KKR calculations confirmed that the ASA is a good approximation for calculating the electronic structure of Fe surfaces, as long as one is not interested in surface states that are relevant, e.g., in scanning tunneling microscopy [110]. It appears therefore that the presence of Friedel-like oscillations in μ_{spin} below crystal surfaces is well confirmed.

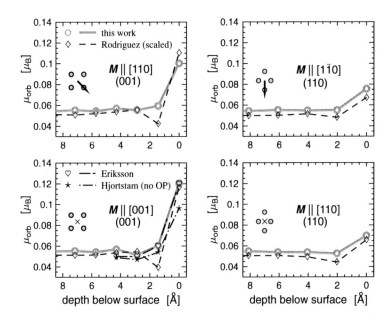

Fig. 3.5. Layer-dependence of μ_{orb} below iron crystal surfaces; left panels stand for the (001) surface, right panels for the (110) surface. The magnetization $\vec{M}$ is either perpendicular (lower panels) or parallel (upper panels) to the surface, its orientation is schematically depicted in the insets. Moments calculated by Rodríguez-López et al. [96] (scaled by a factor of 0.57) and in the case of the (001) surface with perpendicular magnetization also by Eriksson et al. [97] and by Hjortstam et al. [91] (without OP) are shown for comparison.

In Fig. 3.5 we compare μ_{orb} at the (001) and (110) surfaces for the magnetization oriented either perpendicular or parallel to the surface, as computed by the fully-relativistic TB-KKR approach (present work) and by the d-band model Hamiltonian calculations of Rodríguez-López et al. [96]. In the case of the (001) surface with perpendicular magnetization, LMTO calculations of Eriksson et al. [97] and Hjortstam et al.

[91] are also displayed in the graph. The results of Eriksson *et al.* [97] nearly coincide with our results, hence the corresponding line is hard to distinguish in the lower left panel.

Our calculations do not include the so-called orbital polarization (OP) term that was introduced by Brooks to account for the enhancement of spin-orbit induced μ_{orb} due to electronic correlations [111, 112]. As Hjortstam *et al.* [91] present their results both with the OP-term included and without it, we choose for comparison their results obtained without the OP-term. Adding the orbital polarization term to the Hamiltonian is supposed to account heuristically for some many body effects which give rise to the Hund's second rule in atomic theory and which are neglected by the local approximations to relativistic spin density functional theory [35, 113]. These essentially atomic many body effects are presumably better described by the model Hamiltonian approach employed by Rodríguez-López *et al.* [96] or by the LDA+U scheme [114]. In fact the LDA+U as well as the OP-schemes can be combined with the fully-relativistic spin-polarized KKR-formalism used here for the electronic structure calculations [113, 115]. As the present study primarily aims to investigate the dependency of local magnetic properties of Fe surfaces and to compare these with their counterparts for corresponding surfaces of closed-shell Fe clusters, we restricted ourselves to plain LSDA-based calculations. Accordingly, to allow for a comparison of our results for μ_{orb} with those of Rodríguez-López *et al.* [96], we scaled their data down by a factor of 0.57, which is the average ratio between μ_{orb} obtained without and with OP at and below the (001) surface in the work of Hjortstam *et al.* [91].

One can observe in Fig. 3.5 that all calculations provide a very similar enhancement of μ_{orb} at crystal surfaces. Similarly as in the case of μ_{spin}, this enhancement is larger for the (001) surface than for the (110) surface. A good overall agreement between our results and the scaled results of Rodríguez-López *et al.* [96], for the two different surfaces, suggests that the effect of orbital polarization on μ_{orb} at different layers indeed can be roughly estimated by a common multiplicative factor.

A dependence of the orbital moments on the direction of magnetization is pronounced only at the surface. Already in the first subsurface layer the change of the orbital moment is negligible. It gives us a hint that the magnetic anisotropy energy (MAE) of a semi-infinite crystal should be bigger than in the bulk and that the biggest contribution comes from the surface atoms.

3.2.4 Magnetic anisotropy

The experimental results of the MAE for bulk Fe demonstrate the smallness of the anisotropy energy (about 10^{-7} Ry/atom) [2]. As explained in the theoretical overview by Bruno [116], the smallness of the MAE in cubic transition metal bulk is due to the high cubic symmetry and the translational invariance. After breaking the translational invariance at the Fe surface, the MAE increases. As the MAE in bulk cubic transition metals is negligible compared to transition metal surface systems, one would suppose that the Fe layers in the center of a Fe slab do not contribute significantly to the MAE of Fe surface systems. Unlike expected, a strong oscillation of the layer-resolved electronic MAE even in the middle of these thin Fe slabs can be observed in Fig. 3.6. Even though the layer-resolved MAE is not converged in the middle of these thin slabs, the decreasing tendency is visible.

The dipole-dipole contribution increases more or less linearly with the number of Fe layers (shown for the Fe(001) surface in Fig. 3.7). The thicker the Fe film the bigger the dipole-dipole interaction contribution, whereas the electronic contribution is more or less constant for thicker films. From this, one can expect that in the case of very thick Fe slabs the preferable magnetization direction will be in the plane.

3.3 Fe clusters

3.3.1 Structure and computational details

Free spherical-like clusters constructed from 1–7 coordination shells of bulk bcc Fe were investigated by Šipr. The neglect of the geometry relaxation is most serious for small clusters, which in reality may adopt various structures with sometimes tiny differences in their total energies [117, 107]. The structure of larger clusters seems to be less effected by geometry relaxation, as suggested by TB model Hamiltonian calculations [101, 118] as well as by *ab initio* results [108]. In contrast to the previous works, this study primarily focuses on comparing spin and

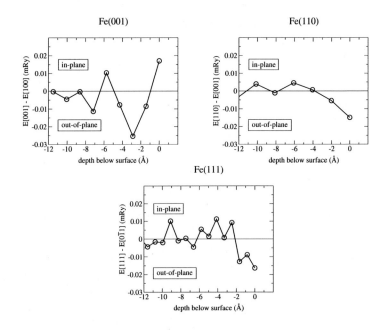

Fig. 3.6. Layer-resolved electronic MAE for the Fe surface systems, given as the difference between an out-of-plane and in-plane magnetization direction.

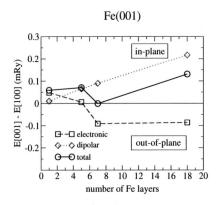

Fig. 3.7. The electronic and the dipole-dipole (dipolar) part of the MAE and the total MAE depending on the thickness of the Fe(001) film.

orbital contributions to the magnetism of surface systems and clusters (in particular relatively large clusters). The neglect of the geometry relaxation for both type of systems seems to be adequate because it allows a more direct comparison.

The electronic and magnetic structure of Fe clusters was calculated in real space via a fully-relativistic spin-polarized multiple scattering technique [35], as implemented in the SPRKKR code [119]. Similarly as in the case of crystal surfaces, the potential was taken in the ASA form. In order to account for the spilling of the electron charge into the vacuum, the clusters were also surrounded by empty spheres. The scattering potential of atoms in the clusters was obtained from scalar-relativistic self-consistent-field (SCF) calculations for clusters using an amended XASCF code [120, 121]. Further technical details on the way of constructing the cluster potential can be found in Ref. [122].

All calculations have been done assuming a collinear spin configuration, i.e., the orientation of the magnetization is characterized by a common vector $\vec{M}$. This restriction might be questionable for the clusters because non-collinear spin structures for the ground state have been indeed found for very small Fe clusters (containing up to five atoms) [123, 124]. Test calculations done for a Fe cluster with 9 atoms using the VASP code [125, 126] lead to a collinear spin configuration [127]. Because also all other clusters studied within this investigations had a highly symmetric cubic geometric structure with closed atomic shells, it is expected that assuming a collinear spin structure is well justified.

3.3.2 Magnetic properties

The dependence of μ_{spin} on the distance of atomic sites from the cluster center is displayed in the left panel of Fig. 3.8, for cluster sizes ranging from 9 atoms (a single coordination shell) to 89 atoms (seven coordination shells around the central atom). In contrast to a non-relativistic or scalar-relativistic description, atoms belonging to the same coordination shell need not be all symmetry equivalent, because the presence of a magnetization and of spin-orbit coupling lowers the symmetry of our systems [128]. Nevertheless, it can be seen in the left panel of Fig. 3.8 that the spin magnetic moments of atoms of the same coordination shell are practically all identical, even if they are inequivalent due to

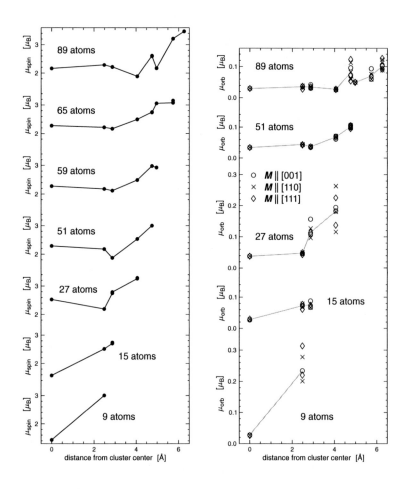

Fig. 3.8. Dependence of μ_{spin} (left panel) and μ_{orb} (right panel) in free iron clusters on the distance of the atoms from the center of the cluster. Each cluster is identified by the number of constituting atoms. In the right panel, the magnetization $\vec{M}$ is oriented parallel to the [001], [110] or [111] crystallographic directions, as indicated in the legend. The solid lines denote μ_{orb} averaged over all atoms in a coordination shell.

the presence of spin-orbit coupling (small differences can be observed only for atoms in the outermost shell of the 65 atoms cluster). Likewise, there is practically no dependence of $\mu_{\rm spin}$ on the direction of the magnetization, in line with previous findings e.g. for bulk systems [61]. This behavior can be explained by the fact that the spin magnetic moment is determined by the difference in the populations of the exchange-split spin up and spin down states, that is hardly changed if the spin-orbit coupling is considered as a perturbation.

On the other hand, $\mu_{\rm orb}$ exhibits quite a strong dependence both on the symmetry class of the atoms within a coordination shell and on the direction of the magnetization. This can be seen in the right panel of Fig. 3.8, where the profiles in $\mu_{\rm orb}$ are shown for three magnetization directions (the magnetization $\vec{M}$ is parallel to one of the [001], [110] and [111] crystallographic directions). The spread in values of $\mu_{\rm orb}$ for inequivalent atoms of the same coordination shell clearly differs from one cluster to another, with no obvious systematic dependence on the cluster size (the spread is small for clusters of 15 and 51 atoms but quite large for clusters of 9, 27, or 89 atoms). Note that the lowering of cluster symmetry induced by the magnetization depends on the direction of $\vec{M}$, meaning that the number of symmetry inequivalent classes into which atoms of the same coordination shell split may be different for different magnetization directions, as it can also be observed in the right panel of Fig. 3.8. In line with this, we find that the spread of $\mu_{\rm orb}$ within an atomic shell is smallest for $\vec{M}$ oriented along the high symmetry direction [001] while it is in general largest for $\vec{M}$ coincident with the direction of the lowest symmetry [110].

It can be deduced from Fig. 3.8 that both $\mu_{\rm spin}$ and $\mu_{\rm orb}$ are enhanced when approaching a cluster surface. This enhancement does not monotonously depend on the distance. Rather, Friedel-like oscillations in $\mu_{\rm spin}$ and $\mu_{\rm orb}$ appear. At some sites, $\mu_{\rm spin}$ or $\mu_{\rm orb}$ may acquire values which are lower than those at the central atom. Convergence of $\mu_{\rm spin}$ and $\mu_{\rm orb}$ to bulk values has not yet been attained even at the center of the 89 atoms cluster (our calculations yield bulk values of $2.28\mu_B$ for $\mu_{\rm spin}$ and $0.054\mu_B$ for $\mu_{\rm orb}$).

Although $\mu_{\rm orb}$ at individual atoms may depend quite strongly on the magnetization direction, $\mu_{\rm orb}$ averaged over all atoms of a given coordination shell (shown via lines in the right panel of Fig. 3.8) do not exhibit any significant dependence on the direction of the magnetic field. This suggests a very small magnetic anisotropy energy (MAE) for closed-shell spherical clusters [84], of the same order as in the bulk. In this respect our clusters differ from smaller low-symmetry clusters investigated by Pastor *et al.* [129], for which a MAE as large as in thin films has been found.

So far we were concerned with magnetization profiles, i.e., with the distribution of μ_{spin} and μ_{orb} within each cluster. However, experiment typically sees only values averaged over all atoms constituting a cluster. Therefore, we present in Tab. 3.1 total μ_{spin} and μ_{orb} of clusters divided by the number of atoms, $\bar{\mu}_{\text{spin}}$, $\bar{\mu}_{\text{orb}}$, and also their ratio $\bar{\mu}_{\text{orb}}/\bar{\mu}_{\text{spin}}$. As the analysis of experimental XMCD spectra on the basis of the sum rules involves the average number of holes in the $3d$ band, this quantity is presented in Tab. 3.1 as well. The Tab. 3.1 reveals that the $\bar{\mu}_{\text{orb}}/\bar{\mu}_{\text{spin}}$ ratio approaches the bulk value much more quickly than $\bar{\mu}_{\text{spin}}$ or $\bar{\mu}_{\text{orb}}$ separately. This might be seen as a contradiction to some experimental XMCD studies which suggest that the $\bar{\mu}_{\text{orb}}/\bar{\mu}_{\text{spin}}$ ratio is about twice as high for supported iron clusters than for the bulk Fe crystal [130–132]. We suppose that one of the main reasons for this discrepancy rests in the shape of the clusters: supported clusters investigated in Refs. [130–132] were probably rather flat than spherical, containing thus a much larger portion of surface and edge atoms with a large μ_{orb} than the spherical clusters we investigate here. A further reason for the different dependency of the $\bar{\mu}_{\text{orb}}/\bar{\mu}_{\text{spin}}$ ratio given in Tab. 3.1 and deduced from the mentioned XMCD investigations is that relativistic calculations based on plain spin density functional theory give the spin-orbit induced μ_{orb} often too small [113] (see also Sec. 3.2.3). This problem is in fact more pronounced for small clusters than for the bulk [133]. Finally, one has to mention that the estimate of $\bar{\mu}_{\text{spin}}$ on the basis of the sum rules is normally based on the assumption that the spin magnetic dipole term T_z in the sum rule can be ignored [134]. Our calculations show that, indeed, for free *spherical* Fe clusters the T_z term is negligible (typically, it is by three orders of magnitude smaller than μ_{spin}). However, for low-dimensional or low-symmetry systems, the contribution coming from T_z may be significant [135].

It is worthwhile to compare our magnetization profiles with earlier

size	$\bar{\mu}_{\text{spin}}$ [μ_B]	$\bar{\mu}_{\text{orb}}$ [μ_B]	$\bar{\mu}_{\text{orb}}/\bar{\mu}_{\text{spin}}$	n_h
9	2.84	0.208	0.0731	2.89
15	2.54	0.070	0.0275	3.02
27	2.83	0.125	0.0441	3.19
51	2.62	0.075	0.0285	3.22
59	2.68	0.062	0.0233	3.27
65	2.66	0.074	0.0281	3.34
89	2.70	0.068	0.0253	3.34
bulk	2.28	0.054	0.0237	3.44

Table 3.1. Magnetic properties of iron clusters averaged over all their atoms as a function of cluster size. The first column displays the number of atoms in a cluster, the second and the third columns show average $\bar{\mu}_{\text{spin}}$ and $\bar{\mu}_{\text{orb}}$, the fourth column contains the ratio of averages $\bar{\mu}_{\text{orb}}/\bar{\mu}_{\text{spin}}$ and the last column shows the average number of holes in the d-band.

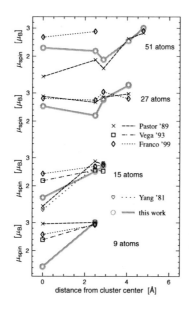

Fig. 3.9. Comparison of μ_{spin} profiles as calculated by different methods. Solid lines correspond to this work, coarsely dotted line to Yang *et al.* [105], dashed line to Pastor *et al.* [101], dash-dotted line to Vega *et al.* [102] and densely dotted line to Franco *et al.* [103]. Each cluster is identified by the number of its atoms.

works. In Fig. 3.9 we display our results together with μ_{spin} obtained
from non-relativistic SCF $X\alpha$ calculations of Yang *et al.* [105] and from
non-relativistic parametrized model calculations of Pastor *et al.* [101],
Vega *et al.* [102] and Franco *et al.* [103]. It is obvious from Fig. 3.9 that
the spin magnetization profiles calculated by different methods show
a rather pronounced spread. The differences are larger for the inner
atoms than for the outer ones. This suggests that even without involv-
ing geometry optimization, the task of calculating electronic structure
of metallic clusters is quite a complex one. On the other hand, in spite
of the rather large quantitative scatter of the various results one nevertheless notices that the qualitative trend of the profiles is in reasonable
agreement.

No other calculations of site-dependent μ_{orb} in free iron clusters have
been published so far, to the best of our knowledge. Guirado-López *et
al.* [109] presented recently results of their TB model Hamiltonian calculation of μ_{orb} in free spherical Ni clusters containing up to 165 atoms.
Although our results for Fe clusters cannot be directly compared with
results for Ni clusters, it is interesting to note that quite a significant
dependence of μ_{orb} averaged over all atoms of a coordination shell on
the magnetization direction was found by these authors, which is in
contrast to the present results for Fe clusters. It is conceivable that
the local geometry of clusters is important in this respect: the Ni clusters investigated by Guirado-López *et al.* [109] have either an fcc or an
icosahedral structure while our Fe clusters have a bcc structure.

3.4   From clusters to surfaces

3.4.1   DOS comparison

We compare the depth profiles of the density of states of the semi-
infinite crystal and the 89 atoms cluster cut along the (001) surface.
To facilitate such a comparison, atomic planes were identified in the
cluster in a way to form subsets of corresponding atomic planes of the
parental crystal; this is shown schematically in Fig. 3.10. As atoms
belonging to the same atomic plane in a cluster are apparently not
all equivalent, the atom most "centrally" located was selected as the
one representing the whole plane (because properties of this atom will

be most resembling the properties of atoms in a semi-infinite crystal plane). Such representing atoms are depicted in black in Fig. 3.10.

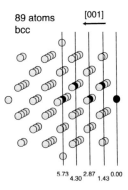

Fig. 3.10. 89 atoms cluster with (001) crystal planes schematically depicted. Numbers below the planes are distances in Å of the respective plane from the "surface plane". Atoms representing each plane are shown in black.

A depth profile of the spin-polarized density of states (DOS) for a semi-infinite crystal and a cluster sliced perpendicular to the [001] direction is presented in Fig. 3.11. The bulk DOS of an iron crystal is shown as well for comparison. The same broadening by a Lorentzian curve with a full width at half maximum of 0.01 Ry was applied to all DOS curves.

One can see that at the surface (d=0.0 Å), both cluster DOS and semi-infinite crystal DOS exhibit more narrow and intensive features than the DOS in the layers below the surface. Especially, the DOS at the cluster surface contains sharp atomic-like-looking features. This is intuitively plausible as this layer actually contains just a single atom (cf. Fig. 3.10).

The convergence towards the bulk case is much slower in the case of a cluster than in the case of a planar surface. Note that the DOS in the third layer below the crystal surface (bottom left panel) is already hardly distinguishable from the bulk DOS. On the other hand, even in the center of the 89 atoms free cluster the DOS still significantly differs from the bulk case.

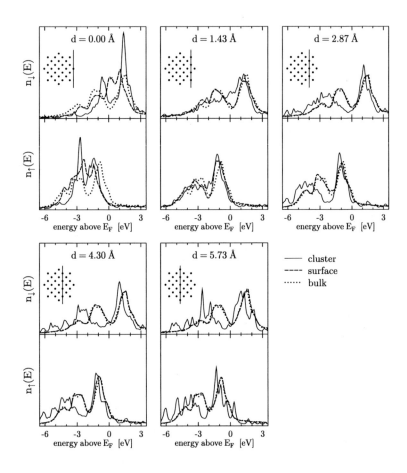

Fig. 3.11. Spin-polarized DOS of atoms in the cluster (full lines) and at the (001) crystal surface (dashed lines). Dotted lines show DOS of the bulk Fe crystal. The distance d from the surface as well as the position of the relevant layer in the 89 atoms cluster are shown in the insets.

Table 3.2. Number of atoms constituting each layer formed when slicing an 89 atoms spherical cluster into planes perpendicular to the [001], [110] and [111] crystallographic directions. The first layer is the outermost one, the last layer is always that one which contains the central atom.

layer	no. of atoms in layer		
	[001]	[110]	[111]
1	1	2	3
2	12	13	1
3	9	18	6
4	16	23	6
5	13		7
6			6
7			12
8			7

3.4.2 Comparison between magnetization profiles of clusters and surfaces

Our calculations of μ_{spin} and μ_{orb} of free clusters and at crystal surfaces were performed within a common theoretical framework, relying on identical or very similar approximations and computational methods. Hence they are well suited for a comparative study of magnetic properties of atoms in free clusters and at crystal surfaces. For this purpose, we focus on an 89 atoms spherical cluster and slice it into atomic layers so that these layers form parts of corresponding planes in the parental bcc crystal. The numbers of atoms in layers perpendicular to three common crystallographic directions are summarized in Tab. 3.2. As atoms belonging to the same crystallographic layer in a cluster are not all equivalent, that atom which is most "centrally" located was selected to represent the whole plane. This choice was made because among all atoms in such a layer, the properties of this atom will resemble most the properties of atoms in the corresponding layer below a crystal surface. At the same time, one has to bear in mind that our comparison of crystal and cluster surfaces concerns the ideal non-relaxed bcc structures. Real clusters will probably have surface faces and interatomic distances different from those of crystal cuts.

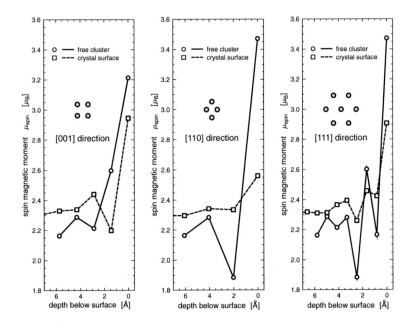

Fig. 3.12. μ_{spin} in an 89 atoms cluster explored in the [001] (left panel), [110] (middle panel) and [111] (right panel) crystallographic directions compared with μ_{spin} at and below corresponding crystal surfaces.

The dependence of μ_{spin} on the distance from the crystal surface and from the surface of a 89 atoms cluster is shown in Fig. 3.12. The enhancement of μ_{spin} at the surface is larger in clusters than in crystals for all three directions we explored, which is consistent with a lower coordination number of the atoms at the surface of a cluster than at the surface of a crystal. The Friedel-like oscillations are also more pronounced in the clusters than for the surface region of crystals. They appear to be in phase for the [110] and [111] directions but not for the [001] direction. This could be intuitively understood given the fact that for semi-infinite crystals, there is only an abrupt termination by the surface in one direction, while for clusters there are many such terminations. The oscillations of μ_{spin} in clusters can thus be viewed as arising from a complex interference of several Friedel-like oscillations. The different phase of the μ_{spin} oscillations for clusters and crystal surfaces in the [001] direction is thus not very surprising.

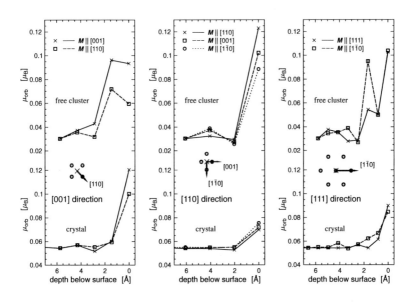

Fig. 3.13. μ_{orb} in an 89 atoms cluster explored in the [001] (left panel), [110] (middle panel) and [111] (right panel) crystallographic directions compared with μ_{orb} at and below corresponding crystal surfaces. The magnetization is either perpendicular to the layers (solid lines) or parallel to them (dashed and dotted lines). For the case of an in-plane magnetization, the direction of the magnetic field $\vec{M}$ is also indicated by the inset drawings. In each panel, the upper graph corresponds to the cluster while the lower graph to the semi-infinite crystal.

Layer-by-layer profiles of μ_{orb} for clusters and semi-infinite crystals are displayed in Fig. 3.13. The magnetization vector $\vec{M}$ is either perpendicular to the layers (i.e., parallel to the direction in which the cluster and the semi-infinite crystal are probed) or parallel to the layers. In the case of the [110] probing direction, different profiles in μ_{orb} are obtained for different in-plane orientations of $\vec{M}$ (middle panel of Fig. 3.13, dashed and dotted lines). For the [001] and [111] probing directions, no such dependence was found (the magnetization profiles for two mutually perpendicular in-plane directions of $\vec{M}$ agree within the thickness of the line).

The surface enhancement of μ_{orb} as well as Friedel-like oscillations are more pronounced at clusters than at surfaces. Similarly, the difference between μ_{orb} for $\vec{M}$ perpendicular to the layers or parallel to them is larger for clusters than at surfaces. Note that this finding does not contradict our earlier statement that there is hardly any anisotropy in *shell-averaged* μ_{orb} (Sec. 3.3.2), because here we focus on μ_{orb} for individual atoms and not on average values.

Like in the case of the DOS, there are some similarities and distinctions in the semi-infinite crystal and cluster cases. Generally, the nearer to the surface the larger are the spin and orbital moments. However, there are also obvious differences in the details of the magnetization profiles of the cluster and the semi-infinite crystal. These differences are more pronounced for μ_{orb} than for μ_{spin}. This is in accordance with an earlier observation that the orbital magnetic moment is especially sensitive to the system geometry [112]. In line with the results for the DOS, the magnetic moments of a semi-infinite crystal converge to the bulk values much faster than magnetic moments of a free cluster. (We obtain $\mu_{spin} = 2.29\mu_B$ and $\mu_{orb} = 0.054\mu_B$ for bulk iron.)

3.4.3 Systematic trends in magnetic moments

The enhancement of μ_{spin} at surfaces is in general ascribed to the reduction of the coordination number of the surface atoms. According to tight-binding considerations, this leads to a narrowing of the electronic band and in turn in general to an increase of the minority DOS at the Fermi level. On the basis of the Stoner criterion one finally expects an increase of μ_{spin} compared to the bulk system. Although isolated clusters – in contrast to a surface regime – have a discrete eigenvalue spectrum, this chain of arguments seem to be applicable for them as well. In fact, it has been shown by many examples that magnetic moments of atoms in transition metals increase if the atomic coordination number decreases [136–138]. However, such a dependence has never been assessed in a quantitative way. Our study incorporates Fe atoms for quite a large range of coordination numbers. Exploring the dependence of μ_{spin} and μ_{orb} on the number of neighbors offers thus a natural way for analyzing the theoretical data we have obtained.

In order to account for the influence of the nearest as well as the next-nearest neighbors, we rely on the effective coordination number N_{eff} (Ref. [139]),

$$N_{eff} = N_1 + \beta N_2 \; , \tag{3.1}$$

where N_1 is the number of the nearest neighbors and N_2 is the number of the next-nearest neighbors. The coefficient β is determined by the distance dependence of the d electron hopping integrals [139,140]. Following Pastor et $al.$ [140] and Zhao et $al.$ [141], we take $\beta = 0.25$, meaning that we have $N_{eff}=9.50$ for atoms in a bulk system with bcc structure. We checked that the main conclusions drawn in this section are not very sensitive to the particular value of the β coefficient. Fig. 3.14 summarizes μ_{spin} and μ_{orb} for all the cluster sizes and crystal surface types we explored as a function of N_{eff}. For clusters we consider μ_{orb} averaged over all atoms in a given shell (which are essentially independent of $\vec{M}$, cf. Sec. 3.3.2), for crystal surfaces we make an average of μ_{orb} for in-plane and perpendicular magnetizations. One can see that μ_{spin} depends on N_{eff} approximately in a linear way, especially if data points for the smallest clusters of 9 and 15 atoms are excluded. Fitting these data in the $N_{eff} < 8.5$ region, we arrive at the following relation

$$\mu_{spin} = -0.21 \times N_{eff} + 3.94 \; . \tag{3.2}$$

If we considered only atoms in clusters, the slope would be a bit more steeper ($\mu_{spin} = -0.22 \times N_{eff} + 3.98$), if we considered only atoms at crystal surfaces, the slope would be more moderate ($\mu_{spin} = -0.16 \times N_{eff} + 3.71$). Such a linear dependence describes $\mu_{spin}(N_{eff})$ only for $N_{eff} \leq 8.0$; for larger N_{eff}, μ_{spin} saturates around the bulk value with considerable deviations of individual data points from this mean value. Note that the linear dependence of μ_{spin} on N_{eff} revealed by Fig. 3.14 differs from the $\sim N_{eff}^{-1/2}$ form which was used in Refs. [141, 142] and which follows from certain assumptions about the character of the DOS and exchange interaction (rectangular d-band, second moment approximation, d-band splitting caused by exchange interaction same for clusters and bulk).

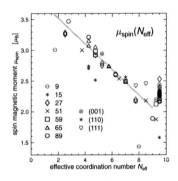
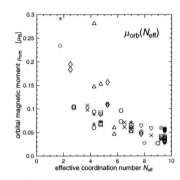

Fig. 3.14. μ_{spin} and μ_{orb} of atoms in clusters and at crystal surfaces as a function of the effective coordination number N_{eff}. Assignment of marks to different clusters and crystal surfaces is indicated by the legend in the left panel. The straight line in the left panel is a fit to the data in the region $N_{\text{eff}} < 8.5$, with the 9 and 15 atoms clusters omitted.

Similarly to the case of μ_{spin}, one can also observe an essentially monotonous decrease of μ_{orb} with increasing N_{eff} for both free clusters and crystal surfaces (right panel of Fig. 3.14). However, one cannot describe this with a mathematically simple relationship as in Eq. (3.2) and also the spread of the values of μ_{orb} for a given N_{eff} is relatively large. Despite this fact, the correlation between μ_{orb} and N_{eff} is obvious, although less clear cut than in the case of μ_{spin}. This μ_{orb}–N_{eff} interrelationship can be explained to some extent by an expression for the spin-orbit induced μ_{orb} that is based on perturbation theory and that relates μ_{orb} to the difference of the DOS at the Fermi level for the spin up and spin down components [95]. Obviously, this difference will depend on the coordination number N_{eff} in a similar way as discussed above for the total DOS at the Fermi level.

The number of valence electrons of atoms in the surface region of a cluster or solid will in general be reduced because of the spill-out of electrons into the vacuum region. Due to the pronounced exchange splitting (see Sec. 3.2.2) and the band narrowing discussed above this will primarily affect the minority spin electrons. As a consequence, one may expect at least a monotonous variation of μ_{spin} with the number of valence electron N_{val} for atoms in the surface region. In fact, as it can be seen in the left panel of Fig. 3.15, a nearly linear relationship

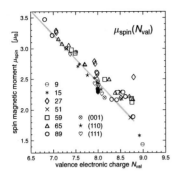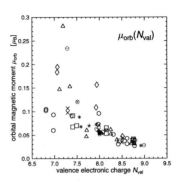

Fig. 3.15. μ_{spin} and μ_{orb} of atoms in clusters and at crystal surfaces as a function of the valence electronic charge N_{val}. Assignment of marks to different clusters and crystal surfaces is indicated by the legend in the left panel. The straight line in the left panel is a fit to the data in the region $N_{\text{val}} < 8.6$.

between μ_{spin} and N_{val} is found when plotting the data for the various systems considered here. In particular one finds, in line with the given arguments, that atoms in the (110), (001) and (111) surface layers of a semi-infinite Fe crystal with effective coordination numbers N_{eff} of 7.00, 5.25, and 4.75 exhibit a reduction of the number of valence electrons ΔN_{val} by -0.23, -0.49, and -0.62 compared to bulk, accompanied by an increase of μ_{spin} by 0.27, 0.65, and 0.61 μ_{B}, respectively. As can be seen from Fig. 3.15, the cluster data – including those for the 9 and 15 atom clusters – follow the trend of the crystal surfaces results. Fitting all the data points satisfying $N_{\text{val}} < 8.6$ by a straight line yields

$$\mu_{\text{spin}} = -0.78 \times N_{\text{val}} + 8.67 \ . \tag{3.3}$$

If one considers data points for clusters only, the slope of the line is more moderate ($\mu_{\text{spin}} = -0.68 \times N_{\text{val}} + 8.04$) while if one takes into account only crystal surfaces, the slope is steeper ($\mu_{\text{spin}} = -1.04 \times N_{\text{val}} + 10.64$). Similarly as in the case of N_{eff}, the dependence of μ_{orb} on N_{val} cannot be described by a simple formula, although a general tendency for increasing μ_{orb} if N_{val} decreases is evident in the right panel of Fig. 3.15.

The systems we study are all bcc-like, with fixed interatomic distances, hence Figs. 3.14 and 3.15 show the net effect of varying the

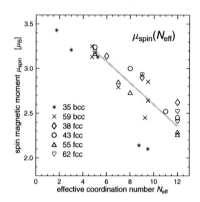

Fig. 3.16. μ_{spin} of atoms in structurally relaxed bcc-like and fcc-like clusters calculated by Postnikov *et al.* [108] displayed as a function of N_{eff}. Assignment of marks to different clusters is indicated by the legend. The straight line is a fit to the data for fcc-like clusters.

coordination numbers. In real free clusters, relaxation of interatomic distances will take place. In order to check to what extent the quasilinear $\mu_{\text{spin}}(N_{\text{eff}})$ dependence holds if variations in distances occur, we focus briefly on μ_{spin} of structurally relaxed free clusters as calculated via an *ab initio* non-relativistic method based on numerical local orbitals in combination with norm-conserving pseudopotentials by Postnikov *et al.* [108]. The clusters considered by these authors were either of bcc type (35 and 59 atoms) or of fcc type (38, 43, 55, and 62 atoms). The corresponding $\mu_{\text{spin}}(N_{\text{eff}})$ dependence for these six clusters is displayed in Fig. 3.16 (the β coefficient of Eq. (3.1) is taken zero for an fcc structure, according to Ref. [141]). It can be seen that the μ_{spin}–N_{eff} interrelationship retains its quasilinear character. Nevertheless, data points for the two bcc clusters noticeably deviate from the pattern set by the four fcc clusters. Data points for the fcc clusters give rise to the approximate relation

$$\mu_{\text{spin}} = -0.12 \times N_{\text{eff}} + 3.81 \, , \tag{3.4}$$

which is also schematically depicted in Fig. 3.16. We can conclude that the approximately linear $\mu_{\text{spin}}(N_{\text{eff}})$ dependence seems to hold for structurally relaxed clusters as well, however, each structure type may have its own "best fit" coefficients.

Chapter 4

FePt surfaces

4.1 Introduction

Alloys comprised of roughly equal proportions of Fe and Pt have been investigated extensively for many years as potential permanent magnets. Below temperatures of 1600K, a $Fe_{0.50}Pt_{0.50}$ alloy undergoes a transition from a fcc based solid solution into an ordered fct ($L1_0$, CuAu structure) phase [143], which has a large value of magnetic anisotropy energy (MAE). Since the large MAE brings about a high thermal stability in magnetic recordings [144], its determination is very important. Films of these Fe-Pt alloys are easy to manufacture and are found to be chemically stable. Recent attention has been paid to high density magnetic recording [145, 146] and magneto-optical recording applications [147]. Since high density recording media with low noise must consist of tiny isolated grains (of size <10 nm^3), the grains need be of a high MAE material so that thermal fluctuations and demagnetizing fields, which could destabilize the magnetization of recorded bits, are avoided. Six years ago, nanocomposite films consisting of magnetically hard FePt nanoparticles in nonmagnetic matrices have been fabricated [148–151]. The FePt particle composition is readily controlled, the particle diameter sizes are tuned between 3 and 10 nm and the nanoparticles can self-assemble into 3D superlattices. Thermal annealing produces $L1_0$-order inside the particles [152] and ferromagnetic nanocrystal assemblies can form. These are chemically and mechanically robust and can support high density magnetization reversal transitions.

Owing to its inherent cubic symmetry, compositionally disordered $Fe_{0.50}Pt_{0.50}$ is magnetically very soft and the alloy only develops large MAE when it orders into the tetragonal CuAu ($L1_0$) structure, i.e., alternately Fe then Pt layers stacked along the [001]-direction. Calculations [63, 68, 153] and measurements [154] suggest potentially the largest MAE for any purely transition metal system with a uniaxial anisotropy constant K_u of the order 10^8 erg/cm^3 [155]. The magnitude of the MAE can be controlled by the extent of the ordering that is induced by annealing the films.

The magnetic anisotropy energy of bulk FePt has been successfully investigated by several theoretical studies [63, 64, 68, 153, 156–158]. The surface-induced properties can strongly differ from the properties of a bulk, as shown in the Chap. 3. The aim of this chapter will be the discussion of the magnetic properties of FePt surface systems. Moreover, the bulk contribution to the total MAE beside the less important surface contribution will be shown.

4.2   Structure and computational details

All FePt systems discussed in this chapter were derived from the FePt binary compound, crystallizing in the CuAu structure. For sake of simplicity the relaxation of the FePt system in the [001]-direction has been omited. The lattice constant for the unrelaxed system (3.807 Å) has been derived from the condition of having the volume of the experimental relaxed system. This lattice constant has been used for all FePt surface systems.

Four different surfaces of FePt have been treated. As shown in the upper panel of Fig. 4.1 the (001) surface can be either Fe- or Pt-terminated. In the case of the (100) surface each layer is occupied by both, the Fe and Pt atoms (see the lower panel of Fig. 4.1). The surface of a completely disordered $Fe_{0.50}Pt_{0.50}$ alloy also consists of both, the Fe and Pt atoms but, on the other hand, they occupy the atomic positions randomly. The influence of disorder on the electronic and magnetic properties of these systems will be investigated in the following.

Describing the surface systems by a slab geometry with a sufficient number of layers ensured a reliable description of bulk-like properties

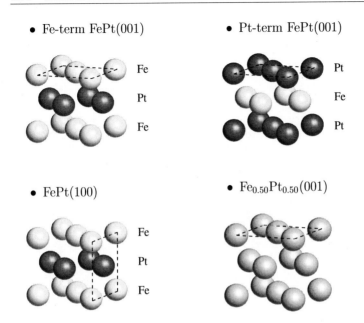

- Fe-term FePt(001)

Fe
Pt
Fe

- Pt-term FePt(001)

Pt
Fe
Pt

- FePt(100)

Fe
Pt
Fe

- $Fe_{0.50}Pt_{0.50}(001)$

Fig. 4.1. Structure of FePt surfaces. Up: Fe-terminated and Pt-terminated (001) surface. Down: FePt(100) surface and $Fe_{0.50}Pt_{0.50}$ (001) surface. Fe atoms are represented by light spheres, while the dark spheres represent Pt atoms. In the case of the disordered $Fe_{0.50}Pt_{0.50}$ alloy each atomic site consists of a mixture of Fe and Pt atoms. The dashed lines denote the surface plane.

in the center of the slabs. In order to account for the spilling of electron charge into the vacuum, all slabs have been surrounded by layers of empty spheres. The electronic and magnetic structure of these slabs has been calculated via the fully-relativistic spin-polarized tight-binding Korringa-Kohn-Rostoker (TB-KKR) method [51, 53] described in Sec. 2.3. We relied on spherical potentials in the atomic sphere approximation (ASA).

4.3 Electronic properties

4.3.1 Bulk FePt

The density of states (DOS) of the bulk FePt is presented in Fig. 4.2. The majority states of Fe atoms (see left panel of Fig. 4.2) are almost completely occupied and because of the big exchange splitting a high spin magnetic moment at Fe atoms can be expected. In the right panel of Fig. 4.2 the large band width of the $5d$ shell of Pt atoms can be observed. The small exchange splitting between the majority and minority states will lead to a small induced spin magnetic moment on Pt atoms.

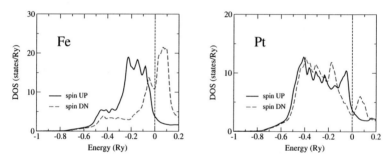

Fig. 4.2. Densities of states of Fe atoms (left panel) and Pt atoms (right panel) in bulk FePt. The solid line represents the majority states, the dashed line the minority states.

4.3.2 Fe- and Pt-terminated FePt(001) surfaces

The layer- and atom-resolved DOS of the Fe-terminated FePt(001) system is shown in upper panels of Fig. 4.3. Fe atoms in the surface layer have only 8 nearest neighbors instead of 12 in the bulk system. Because of the smaller number of nearest neighbors, a sharpening of the DOS in the Fe surface layer (L=0 in left upper panel of Fig. 4.3) has been observed. On the other hand, the DOS curves of the next Fe and Pt layer under the surface (L=2 and 3) show bulk-like features (compare with Fig. 4.2). It means that the effect of the surface is of short-range character, influencing only few layers at the surface. The equivalence

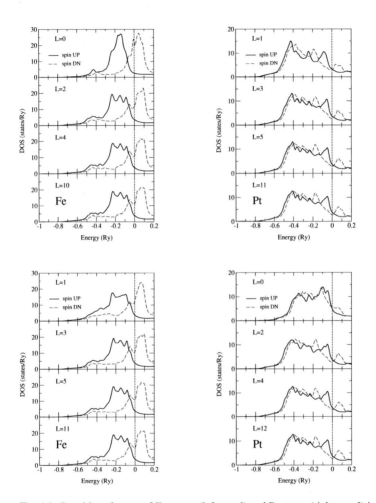

Fig. 4.3. Densities of states of Fe atoms (left panel) and Pt atoms (right panel) in the Fe-(upper panels) and Pt-terminated (lower panels) FePt(001) system. The surface layer is denoted by "L=0", the subsurface layer by "L=1", the underlying layer by "L=2" and so on. The lowest graphs in the columns show the layers in the center of the slab. The solid line represents the majority states, the dashed line the minority states.

of the DOS curves from central Fe and Pt layers (L=10 and 11) with the DOS of bulk FePt confirms the sufficient number of layers in the slab calculation.

The DOS of the Pt-terminated FePt(001) system is shown in lower panels of Fig. 4.3. Due to the larger band width of the $5d$ shell (Pt) in comparison with the $3d$ shell (Fe) the DOS of the Pt surface atoms is not becoming as narrow as the DOS of the Fe surface atoms at the Fe-terminated surface.

4.3.3   FePt(100) and disordered $Fe_{0.50}Pt_{0.50}$(001) surface

The layer- and atom-resolved DOS of the FePt(100) system is presented in upper panels of Fig. 4.4. As already mentioned, each layer of the FePt(100) system consists of Fe and Pt atoms. Like in the previous layered systems, the DOS of the surface atoms is sharpened. A bulk-like DOS is achieved already in the second subsurface layer (L=2).

To describe a disordered $Fe_{0.50}Pt_{0.50}$ alloy the CPA method (see the subsection 2.2.4) has been used. The narrowing of the DOS near the surface in the $Fe_{0.50}Pt_{0.50}$ surface system is shown in lower panels of Fig. 4.4. The DOS of the bulk-like central layers of the disordered $Fe_{0.50}Pt_{0.50}$ surface system (L=12 in lower panels of Fig. 4.4) is almost identical to that reported previously for the bulk alloy $Fe_{0.50}Pt_{0.50}$ [159]. In the FePt(100) and disordered $Fe_{0.50}Pt_{0.50}$(001) surface systems there are Fe and Pt atoms present in each layer, therefore their DOS curves look quite similar. Anyway, in the disordered system there are contributions from all possible structural configurations included. Therefore, the DOS of the disordered alloy is less structured than in the three investigated ordered systems.

4.4   Magnetic properties

4.4.1   Magnetic moments

Investigating the DOS of the bulk FePt (see Fig. 4.2) there is big exchange splitting of the majority and minority states at the Fe atoms with the majority states fully occupied. One can assume that the Fe atoms act as a "source of magnetism" with a quite high spin moment

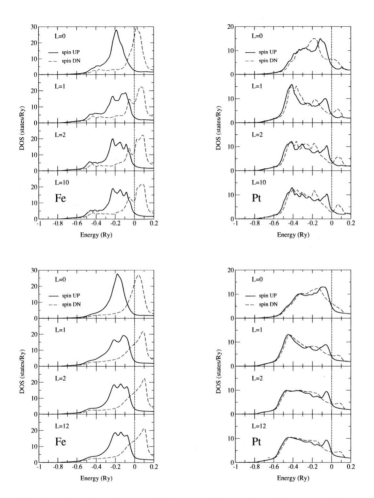

Fig. 4.4. Densities of states of Fe atoms (left panel) and Pt atoms (right panel) in the FePt(100) (upper panels) and in the disordered $Fe_{0.50}Pt_{0.50}(001)$ (lower panels) system. The surface layer is denoted by "L=0", the first subsurface layer by "L=1", the second subsurface layer by "L=2". The lowest graphs in the columns show the layers in the center of the slab. The solid line represents the majority states, the dashed line the minority states.

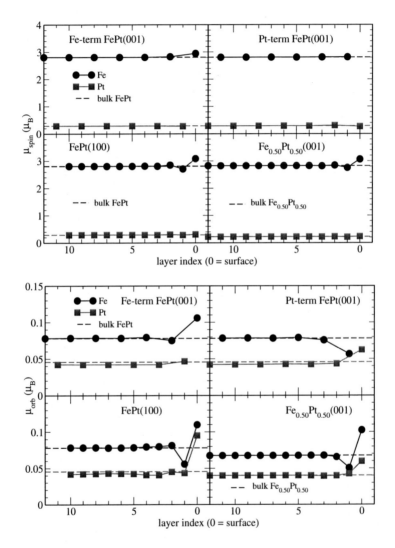

Fig. 4.5. Spin and orbital moment profiles of Fe and Pt atoms. The magnetization direction is parallel to the $\hat{c}_{CuAu}$ axis. The dashed lines give the corresponding values of bulk FePt.

of 2.81 μ_B (2.77 μ_B in the theoretical work of Solovyev *et al.* [68]). The Fe atoms induce the spin magnetization at the Pt sites of 0.29 μ_B (0.35 μ_B by Solovyev *et al.* [68]). Due to the high spin-orbit coupling of Pt atoms the spin magnetization gives rise to a Pt orbital moment of the same magnitude as at the Fe sites (dashed lines in lower panel of Fig. 4.5). Orbital moments obtained by pseudoperturbation treatment of the spin-orbit interaction by Solovyev *et al.* (0.074 μ_B for Fe atoms and 0.044 μ_B for Pt atoms in FePt compounds) coincide with our results within few percents.

In the upper panel of Fig. 4.5 the spin moment profiles of FePt surfaces are presented. In the surface layers the DOS is sharpened and a slight increase of the Fe spin moment is expected. The spin magnetization profiles show a monotonous behavior in the layers below surface. The induced spin magnetic moment of Pt does not deviate significantly from the bulk value throughout the whole slab.

The orbital magnetic moments in FePt surface systems are more sensitive to the depth of the atoms below the surface (see lower panel of Fig. 4.5). At the surface layer of all systems an increase of the orbital moment can be seen. Oscillations of the orbital magnetization profile in the FePt(100) and the disordered $Fe_{0.50}Pt_{0.50}(001)$ surface systems are more pronounced than for the spin magnetization profiles. The magnetization direction was chosen perpendicular to the layers of the CuAu structure. This is indeed the preferable magnetization direction in the ordered FePt surface systems, as will be shown in the next subsection.

4.4.2 Magnetic anisotropy

The magnetic anisotropy energy (MAE) for ordered FePt systems will be defined as follows

$$\Delta E = E(\vec{M} \parallel \hat{c}_{CuAu}) - E(\vec{M} \perp \hat{c}_{CuAu}) \; , \qquad (4.1)$$

where the $\hat{c}_{CuAu}$ is the direction perpendicular to the layers of the CuAu structure (see Fig. 4.1 for the CuAu structure). $\hat{c}_{CuAu}$ is also the preferable magnetization direction of bulk FePt found in experiments [154, 155]. *Ab initio* calculations find the MAE of completely $L1_0$-ordered relaxed bulk FePt (-0.258 mRy [63, 64], -0.250 mRy [68] and -0.300 mRy [153]) in a quite good agreement with the present work (-0.177 mRy). Estimates of the uniaxial MAE of ordered FePt in excess of -10^8 erg/cm^3 (-0.13 mRy per formula unit) have been deduced from measurements on thin films [154, 155] at room temperature. In comparison with the experimental data, the values for the MAE obtained in the theoretical works are overestimated. One of the origins of the discrepancies could be the orbital polarization effects: exchange and correlations between electrons related with orbital degrees of freedom which are essentially beyond the uniform electron gas based LSDA description employed in the present work. Also the temperature dependence of the MAE could be very important, because the present band structure calculations correspond to $T=0$K, whereas all measurements have been performed at finite temperatures. Recently, Staunton et al. [158] succeeded to calculate the temperature dependent magnetic anisotropy of the $L1_0$-ordered FePt system.

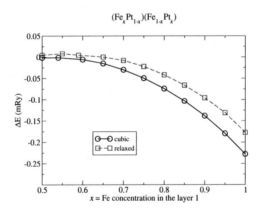

Fig. 4.6. MAE for the (Fe$_{0.50}$Pt$_{0.50}$ – FePt) bulk system dependent on the local disorder.

It is also very likely that the experimental samples were incompletely ordered. Ostanin et al. [157] showed that the degree of ordering strongly affects the MAE and only the $L1_0$-ordered FePt shows the high MAE. The influence of the local disorder on the MAE of bulk FePt can be investigated by using the CPA alloy theory and putting a mixture of 90% Fe and 10% Pt on the layer 1 and a mixture of 90% Pt and

10% Fe on the layer 2 (corresponding to $x=0.9$ in Fig. 4.6). A completely disordered cubic $Fe_{0.50}Pt_{0.50}$ has by definition no MAE ($x=0.5$ in Fig. 4.6). To produce a large MAE, this complete disorder has to be replaced by the $L1_0$- (CuAu)-type compositional ordering with its structure of alternating Fe and Pt layers ($x=1$ in Fig. 4.6). With increasing local ordering the absolute value of MAE increases smoothly. As the structure of the bulk FePt is relaxed and the structure used in following calculations for the FePt surface systems is cubic (unrelaxed), the influence of the local disorder for both – relaxed and cubic – structures is shown in Fig. 4.6. An interesting feature is that the MAE in the relaxed ordered structure is lower by 0.05 mRy, whereas in the completely disordered case both systems show no magnetic anisotropy.

The MAE has a negative sign for all ordered FePt surface systems and the preferable magnetization direction is perpendicular to the layers of the CuAu structure, just like in the bulk FePt. So the preferable magnetization direction is independent of the FePt surface orientation. The contribution to the MAE from the corresponding layers is shown in Fig. 4.7. In the center of the slab the MAE is constant, only near the surface abrupt changes are visible. It follows that the surface contribution does not play an important role in a semi-infinite FePt system because of the dominating contribution from many layers in the center. The dipole-dipole contribution is one magnitude smaller than the electronic MAE and can be neglected for these systems.

In the case of a fcc $Fe_{0.50}Pt_{0.50}(001)$ surface system there is a negative electronic contribution to the MAE. Analogously to the bcc Fe planar surfaces (see the subsection 3.2.4), the dipole-dipole contribution compensates the negative sign of the electronic part and the preferable magnetization direction in the completely disordered $Fe_{0.50}Pt_{0.50}$ surface system is in the plane.

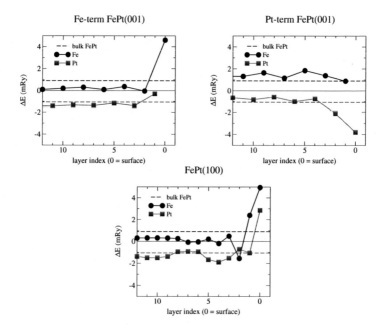

Fig. 4.7. Layer-resolved MAE for the FePt surface systems, given as the difference between an out-of-plane and in-plane magnetization direction.

Chapter 5

GaAs/Fe heterogeneous systems

5.1 Introduction

Spin- or magnetoelectronics, that exploits besides the charge also the spin of the electrons as an additional degree of freedom, receives a lot of interest recently because of its great technological potential [5]. The most prominent example for this is the giant magnetoresistance (GMR) [160] that can be observed in layered or granular ferromagnet/conductor/ferromagnet systems and is used with great success for example in read heads of hard disks. Another promising effect is the tunneling magnetoresistance (TMR) that denotes the observation that the tunneling current across a tunneling barrier that separates two magnetic layers depends on the relative orientation of the magnetization of these layers. This effect, first predicted by Jullière [161], could be demonstrated several years ago for a layered ferromagnet/insulator/ferromagnet (FM/I/FM) system by Moodera et al. [162] and first prototypes of a dynamic storage chip could be presented. Devices based on ferromagnet/semiconductor/ferromagnet (FM/SC/FM) heterostructures seem to have several advantages compared to insulator-based tunneling junctions. Moreover, recent experiments [163–165] could show that a measurable spin-polarized injection can be obtained for a SC/FM system even at room temperature.

Corresponding theoretical work started already in the 80-ies by Pickett and Papaconstantopoulos [166] who performed tight-binding calculations for the Ge(110)/Fe interface. Most of the later work concentrated on multilayer systems based on bcc Fe because this ferromagnetic metal has a structure that nearly perfectly matches to Ge, GaAs and ZnSe substrates [167, 168]. While the first theoretical investigations on these heterogeneous systems were aiming at their electronic structure and magnetization profiles [167, 169–174], more recent work concentrates on the transport properties [175–184]. Nevertheless, a firm basis for the understanding of the transport properties of the FM/SC/FM systems cannot be achieved without a full understanding of their electronic and magnetic properties. This route is followed in Sec. 5.3 by studying SC/FM multilayer systems using standard band structure techniques. Because the electronic and magnetic properties at the interface of SC/FM systems are of central importance for the transport properties, special emphasis will be laid on these. In particular, the spin and orbital magnetization profiles will be presented.

Most important in this context is of course the occurrence of magnetically dead FM layers at the interface and the presence of an induced magnetization in the SC subsystem. Controversial discussions in the literature suggest the occurrence of magnetically dead layers due to the interdiffusion at the GaAs–Fe interface. The effect of interdiffusion at the interface has been studied by means of Coherent Potential Approximation (CPA). Finally, it is suggested that the X-ray magnetic circular dichroism (XMCD) should supply a suitable tool to investigate the former issue, e.g. the induced magnetization in semiconductors. Corresponding spectra for Ga and As in GaAs/Fe will be presented.

Apart from the transport properties of a FM/SC/FM TMR-system, its magnetic anisotropy properties are of central importance for its operation. Accordingly, there is a large number of experimental investigations that dealt with the magnetic anisotropy of layered SC/FM systems [185–190], with a strong emphasize on Fe films on (001) oriented GaAs [191–204]. Depending on the preparation conditions the clean (001) surface of Ga-terminated GaAs undergoes a 4×2- or 2×6- reconstruction [205]. In both cases, however, an uniaxial twofold in-plane anisotropy is found for deposited thin Fe films that have their easy (hard) axis along the [110]- ([1$\bar{1}$0]-) axis of the underlying GaAs substrate. In line with the bcc structure of Fe a biaxial fourfold in-plane anisotropy is found for thicker films [195].

All calculations of the magnetic anisotropy energy (MAE) made so far for SC/FM systems were restricted to multilayers. Using the

fully-relativistic LMTO method Cabria *et al.* [79] found a perpendicular anisotropy for the investigated 5GaAs/xFe (x=1, 3, 5, 7 and 9)
multilayers. Sjöstedt *et al.* [80] could demonstrate the uniaxial in-plane
anisotropy in the case of ZnSe/Fe. In contrast to these previous theoretical studies, we present a detailed study of thin Fe films on (001)
oriented GaAs surface in Sec. 5.4.

5.2 Studied systems

For the calculations to be presented in Sec. 5.3, the SC/FM multilayers
forming a superlattice with the Fe layers perfectly matching to the
GaAs layers have been assumed (see Fig. 5.1). For the zinc-blende
structure of the GaAs layers the lattice parameter $a_{\mathrm{GaAs}} = 5.65$ Å has
been used. According to this, the lattice parameter a_{Fe} for the bcc
structured Fe layers was set to 2.825 Å. In the GaAs layers, so-called
empty spheres have been introduced to cover the open space between
the atoms. The assumption of a periodic superlattice allows one to

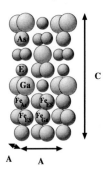

Fig. 5.1. Unit cell of 5GaAs/3Fe multilayer with $A = \sqrt{2}/2 \, a_{\mathrm{GaAs}}$ and $C = 2a_{\mathrm{GaAs}} = 4a_{\mathrm{Fe}}$.
The atomic layers are oriented horizontally. The label E denotes an empty sphere, while
$\mathrm{Fe_{As}}$ and $\mathrm{Fe_{ES}}$ denote the two inequivalent Fe atoms at the SC–FM interface. $\mathrm{Fe_{As}}$ ($\mathrm{Fe_{ES}}$)
indicates that the Fe sites would be occupied by an As atom (empty sphere) if the GaAs
layers continued beyond the interface.

use standard band structure schemes to calculate the corresponding
electronic structure on the basis of the standard spin-polarized KKR
method developed by Akai [206].

To calculate the magnetic anisotropy energy for thin Fe films on top of a GaAs substrate one has to use a slab geometry with two-dimensional periodicity and the spin-orbit coupling has to be included. Therefore a fully-relativistic spin-polarized version of the TB-KKR method has been developed by Popescu in our working group in collaboration with the Jülich group [51, 53, 207]. The substrate was represented by a stack of 3 unit cells of GaAs (13 GaAs layers). For the unit cells we used the zinc-blende structure and the lattice parameter of bulk GaAs (a=5.65 Å). The atomic sphere approximation (ASA) has been used for the potential and charge geometry, with the space between the Ga and As atoms filled by two empty spheres per bulk unit cell. On top of the Ga-terminated GaAs layer we added x Fe layers (x=1–7) having the bcc structure and its lattice parameter adjusted to that of GaAs. A perfect interface without interdiffusion or layer relaxation has been assumed throughout. This geometry is shown in Fig. 5.2 for 4 Fe layers on top of GaAs substrate together with a detailed description of the interface layers. The transition to the vacuum regime, finally, has been represented by 3 layers of empty spheres.

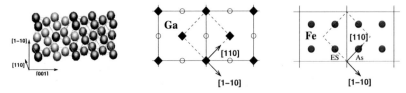

Fig. 5.2. The left panel shows the structure of the investigated GaAs/Fe systems. In the middle panel the terminating Ga layer (Ga: full diamonds, empty spheres: open circles), in the right panel its adjacent Fe layer is shown. The labels ES and As indicate that the Fe sites would be occupied by an empty sphere or an As atom, respectively, if the GaAs layers continued beyond the interface.

5.3   GaAs/Fe multilayers

5.3.1   Spin magnetization profile

In order to get a picture of the magnetic ordering at the interface of SC/FM layered system we will examine the distribution of the magnetic

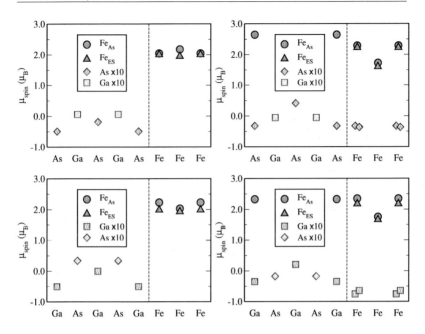

Fig. 5.3. Spin magnetization profile for periodic As-terminated (upper panels) and Ga-terminated (lower panels) 5GaAs/3Fe multilayers. The left panels show results for ordered multilayer systems, whereas the right panels show results for an alloy formation at the interface with 50% of the As (Ga) and Fe atoms of the neighboring layers randomly interchanged.

moments for various atomic layers. For the GaAs/Fe system corresponding theoretical investigations have been performed for GaAs layers sandwiched by half-infinite bcc Fe [167] and also for periodic GaAs/Fe multilayer systems [78]. Within the standard scalar-relativistic KKR method we have calculated the spin magnetic moments for As- and Ga-terminated 5GaAs/3Fe multilayer systems. Fig. 5.3 shows in the left panels the corresponding results for the perfectly ordered systems. As for all previous theoretical studies, the Fe spin moments for the individual layers differ only slightly from that of pure bcc Fe. One can note that there are small variations from layer to layer but the differences between moments for the two inequivalent Fe atoms within a layer are in the same order of magnitude. Most important, however, is that there are no magnetically dead layers with a vanishing or strongly suppressed moment found for the Fe interface layers.

These results seem to be in contradiction with some experimental investigations for Fe grown on GaAs(001), that showed that Fe layers at the interface should have a small or vanishing moment [208]. This was ascribed to the formation of Fe_xAs_y compounds in the interface region at the temperature of 500^oC. Antiferromagnetic ordering of Fe_2As with magnetic moments 2.05 and 1.28 μ_B for the two inequivalent Fe atoms has been observed in experiment [209] (our calculation shows magnetic moments of 2.19 and 1.03 μ_B). FeAs has been reported to be a heli-magnet with a double spin spiral arrangement whose Fe moments are 0.5 μ_B and $FeAs_2$ is a diamagnetic semiconductor [210]. It is obvious that magnetic properties of these compounds could explain the loss of magnetization at the GaAs–Fe interface.

We considered the same As- and Ga-terminated 5GaAs/3Fe multi-layer systems as above to investigate the effect of interdiffusion at the interface (see right panels in Fig. 5.3). For this purpose, we modeled an interchange of Fe and As (Ga) atoms at the interface and performed scalar-relativistic KKR-CPA calculations. The results for an alloy for-mation at the interface with 50% of the As (Ga) and Fe atoms of the neighboring layers randomly interchanged are plotted in the right pan-els of Fig. 5.3. As can be seen this has quite pronounced influence on the magnetization profile. First of all, one notes that the Fe spin moments for the two interface layers are not suppressed but increased, while that of the middle Fe layer decreased. Altogether, a slight in-crease of the total magnetic moment results from these changes. In the two upper graphs of Fig. 5.4 we are showing the behavior of the spin moments of Fe and SC atoms in the Ga-terminated system with increasing interdiffusion. Although a diffusion of As and Ga into the Fe overlayer is well known [211, 212], the calculated total energy values for different degrees of interdiffusion state that the most stable system should be the perfectly ordered one without interdiffusion (see lower panel of Fig. 5.4). Moreover, Fe films deposited by a temperature less than 200^oC on As-terminated [191], Ga-terminated [192, 213] and pas-sivated GaAs substrates all display similar magnetic properties. The 3d charge transfer seems to be weakly affected by the termination of the

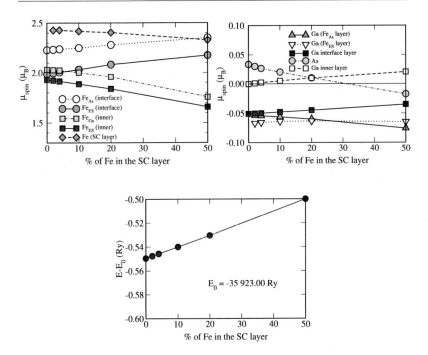

Fig. 5.4. Spin magnetization profile of Fe (left) and SC atoms (right) for Ga-terminated 5GaAs/3Fe multilayer system after inclusion of interdiffusion. The influence of interdiffusion on the total energy is shown in the lower panel.

substrate and by the resulting difference of intermixing at the interface as well [214].

Finally, one has to point out that in all calculations on SC/FM layered systems done so far, a non-vanishing induced magnetic moment for the SC subsystem was found. In all cases the induced moments are highest at the interface. Furthermore, these change in general their sign from layer to layer.

5.3.2 XMCD investigations

Most of the experimental investigations on SC/FM layered systems performed so far have been done by global magnetometer measurements. In particular the presence of so-called magnetically dead Fe layers at the SC–FM interface has been deduced until now only indirectly by measuring the total magnetization of SC/FM bilayer systems as a function of the FM layer thickness [215, 216]. In a recent Mößbauer experiment [217] depositing approximately 2.5 monolayers of Mößbauer-active Fe isotopes at the Ga-terminated interface there were no magnetically dead layers found and a slight interdiffusion of Ga into the Fe layers was observed. Corresponding NMR experiments on ^{69}Ga, ^{71}Ga or ^{75}As would also allow to get corresponding information for the SC layers. In particular, these experiments could confirm the induced spin-polarization in an element resolved way. Anyway, the injection of spin-polarized electrons from a metal into a semiconductor was already demonstrated by Zhu *et al.* [163] for a GaAs/(In,Ga)As light emitting diode covered with Fe. The circular polarization degree of the observed electroluminescence revealed a spin injection efficiency of 2%.

An alternative is to perform X-ray magnetic circular dichroism (XMCD) experiments that also supply element specific information on the magnetic properties of the absorbing atoms. First of all, they may provide layer-resolved information on the investigated system, e.g. by varying the layer sequence [218], by making use of resonant scattering [219] or reflectometry [220]. More important seems to be the fact that the so-called sum rules [134, 221] allow one to deduce a rather reliable estimate for the spin and orbital magnetic moments of the absorbing atoms. Finally, by extending the investigated energy range of the measurements to the EXAFS region, i.e. to several hundreds eV's above the absorption edge, one obtains information on the structural as well as magnetic properties of the atomic region around the absorbing atom.

X-ray magnetic circular dichroism (XMCD) denotes the observation that for magnetic materials the absorption coefficients μ_λ for left and right circularly polarized X-rays differ [222, 223]. To motivate corresponding experiments for the SC/FM systems studied here, the absorption coefficients for Ga and As in GaAs/Fe have been calculated. Using Fermi's golden rule the X-ray absorption coefficients $\mu^{\vec{q}_p \lambda}(\omega)$ can be expressed by [223]:

$$\mu^{\vec{q}_p\lambda}(\omega) \propto \sum_{i\,\text{occ}} \langle \Phi_i | X^{\times}_{\vec{q}_p\lambda} \operatorname{Im}G^+(E_i + \hbar\omega) X_{\vec{q}_p\lambda} | \Phi_i \rangle\, \theta(E_i + \hbar\omega - E_F) \quad (5.1)$$

where $\vec{q}_p$, ω and λ stand for the wave vector, frequency and polarization of the radiation with $X_{\vec{q}_p\lambda}$ being the corresponding electron-photon interaction operator [223]. The sum runs over all involved core states Φ_i. The itinerant final states above the Fermi level are represented by the electronic Green's function obtained within multiple scattering theory (see Sec. 2.2). Because the XMCD is caused by the interplay of magnetization and spin-orbit coupling, the fully-relativistic version of the SPR-KKR cluster method [224] has been used to deal with the expression for μ_λ given in Eq. (5.1), using the self-consistent potentials previously obtained from a fully-relativistic SPR-LMTO calculation for the multilayer system.

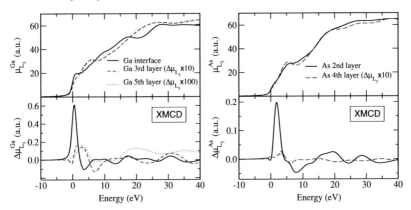

Fig. 5.5. Calculated X-ray absorption coefficient for unpolarized radiation $\bar{\mu}_{L_3}$ for Ga (left) and As (right) in the Ga-terminated 9GaAs/5Fe multilayer systems. The additional spectra show the magnetic circular dichroism, i.e. $\Delta\mu_{L_3}$ gives the difference in absorption for left and right circularly polarized X-rays.

To stimulate experimental XMCD investigations on the SC/FM systems we have calculated the L_3-edges spectra for Ga and As in the Ga-terminated 9GaAs/5Fe multilayer system for near edge region (XANES) using Eq. (5.1). The resulting layer-projected spectra $\bar{\mu}_{L_3}$ for unpolarized radiation are shown in Fig. 5.5 together with the corresponding dichroic spectrum $\Delta\mu_{L_3}$, that gives the difference in absorption for left and right circularly polarized radiation. First of all one has to note that $\Delta\mu_{L_3}$ is non-vanishing if there is a finite magnetic moment present for the absorbing atom – more strictly spoken $\Delta\mu_{L_3}$ reflects the spin and orbital polarization, $\frac{d}{dE}\langle\sigma_z\rangle_d$ and $\frac{d}{dE}\langle l_z\rangle_d$, respectively, of the absorbing atom [224]. As can be seen, Ga atoms at the SC–FM interface have the largest induced moment and consequently the largest $\Delta\mu_{L_3}$ contribution to the dichroic spectra (left panel of Fig. 5.5). Although the corresponding dichroic signal $\Delta\mu_{L_3}$ seems to be quite small compared with $\bar{\mu}_{L_3}$, it is of the same order of magnitude as for example that found experimentally for Cu in Co/Cu multilayer systems [225]. For this reason there is no doubt that it should be possible to study the XMCD of Ga in GaAs/Fe by experiment. According to the smaller moment predicted for the second inequivalent Ga layer in 9GaAs/5Fe system (in the 3rd layer from the GaAs–Fe interface) one can expect a smaller dichroic signal $\Delta\mu_{L_3}$ for these. Of course, in an experiment the dichroic spectra of all the Ga layers would be superimposed. But varying the layer sequence, for example, it is often possible to deduce the individual contributions of the various layers.

Fig. 5.3 shows that the calculations also predict a small but nevertheless finite induced spin magnetic moment for the atoms in the second layer from the GaAs–Fe interface, it means for the As atoms in the case of Ga-terminated system. This should of course be also reflected by a corresponding XMCD signal at the As L_3-edge, as shown in the right panel of Fig. 5.5. Again this signal is quite weak (about 30% less than the $\Delta\mu_{L_3}$ of Ga) but, nevertheless, it should be possible to detect it in experiment.

Confirmation of the predicted induced magnetization in the SC subsystems of SC/FM layer systems by XMCD or related experiments seems to be very important to support the various conclusions drawn on the basis of corresponding band structure calculations. More important, however, the additional proofs for the presence of an induced spin-polarization by experimental techniques would have far reaching consequences for the understanding of the transport properties of SC/FM layer systems [175].

5.4 Fe films on GaAs (GaAs/nFe systems, n=1–7)

5.4.1 Spin and orbital magnetization profile

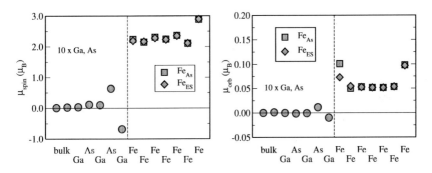

Fig. 5.6. Spin and orbital magnetization profile of the GaAs/7Fe system. The orbital moments in the right panel have been calculated for the [110] magnetization direction (easy axis). The negligible contributions of the empty spheres and vacuum region have been omitted.

The left panel of Fig. 5.6 shows the spin magnetization profile of the GaAs/7Fe system. The main features of the spin magnetization profile at the interface are similar to those of GaAs/Fe multilayer systems shown in the left panels of Fig. 5.3 in the sense that the spin magnetic moments at the GaAs–Fe interface have bulk-like values. The enhancement of the spin moment in the Fe surface layer appears according to the lower coordination number at the surface and is similar to the enhancement at the free Fe surface in the left panel of Fig. 3.4.

The orbital magnetization profile in the right panel of Fig. 5.6 has been calculated with the orientation of the magnetization along the [110]-direction, which is the easy axis for GaAs/Fe systems. The enhancement at the Fe surface corresponds to the enhancement of the orbital moment at the free Fe surface in left panel of Fig. 3.5. One can see that the different GaAs bonding with the two inequivalent Fe atoms leads to a unequal increase of the orbital moment. The results in Fig. 5.6 show in addition that the interface-specific behavior is restricted to the immediate vicinity of the GaAs–Fe interface and bulk-characteristic values for both components are obtained already in the third or fourth layer away from the interface.

5.4.2   Validity of Bruno's and van der Laan's formulas

In the following results for the MAE for the GaAs/Fe systems will be presented. For a detailed description of methods for determining the MAE and its relations to the orbital moment anisotropy (Bruno's and van der Laan's formulas) see Sec. 2.4. For an application of these models for the thin Fe films on GaAs, we used for all Fe atoms the same value of the spin-orbit coupling parameter ξ_{Fe-d} (4.5 mRy [78]) and ignored contributions from the SC subsystem. Having access to both calculated quantities, MAE and the anisotropy of the orbital moments, one can check directly the validity of the two models.

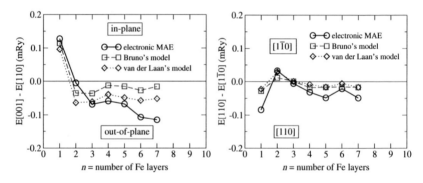

Fig. 5.7.   Electronic part of the MAE for GaAs/nFe systems with n=1–7. The left panel shows the difference between out-of- and in-plane orientation of the magnetization, while the right hand side gives the difference for an in-plane orientation along the [110]- and [1$\bar{1}$0]-directions. In addition the MAE obtained on the basis of Bruno's [84] and van der Laan's [85] expressions is shown.

The results for the electronic part of the MAE obtained for the GaAs/nFe systems with n=1–7 are shown in Fig. 5.7. As one can see, this part of the MAE favors an out-of-plane orientation – apart for the case of a Fe monolayer (n=1). Among the various inequivalent directions parallel to the SC–FM interface, the [110]-direction is favored (right panel of Fig. 5.7). As Fig. 5.7 shows, the directly calculated

MAE is reproduced in a semi-quantitative way by the approximate values obtained on the basis of Bruno's and van der Laan's expressions. Obviously, due to the simplifying assumption on which Bruno's formula is based, it leads to a larger deviation from the directly calculated MAE. Nevertheless, both model calculations clearly demonstrate the close interrelation of the MAE and the anisotropy of the orbital angular momentum. This implies in particular that the later quantity, determined for example by exploiting the magnetic circular dichroism in X-ray absorption, can be used indeed to monitor the MAE [226] of the systems considered here.

5.4.3 Out-of-plane magnetic anisotropy

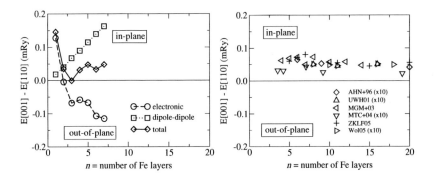

Fig. 5.8. MAE for GaAs/nFe systems with n=1–7, given by the difference between out-of- and in-plane orientation of magnetization. The left panel shows theoretical results and their decomposition into electronic and dipole-dipole part. The right panel shows experimental data in general determined at room temperature. The data have been scaled for that reason by a factor of 10 (see text). References to the experiments: AHN+96 [194], UWH01 [197], MGM+03 [201], MTC+04 [199], ZKLF05 [200] and Wol05 [227].

To compare the results for the MAE with corresponding experimental data the shape anisotropy (dipole-dipole) contribution has to be added to the electronic part. Comparing the out-of- and in-plane orientation of $\vec{M}$ one finds the shape anisotropy to vary nearly linearly with the number of Fe layers for the investigated range of n (see the dipole-dipole contribution in the left panel of Fig. 5.8). As the shape anisotropy compensates the negative electronic part of MAE for n=2–7 an in-plane anisotropy results for all layer thicknesses studied. This is in contrast to the previous work on GaAs/Fe multilayers for which an out-of-plane anisotropy has been found, due to a larger electronic part of the MAE and a smaller shape anisotropy [79]. The right panel of Fig. 5.8 shows that also in a large number of experimental investigations an in-plane anisotropy has been found. While our theoretical data correspond to the temperature T=0K, the experimental measurements were in general done at room temperature. Because the Curie temperatures of thin Fe films decrease steeply with decreasing Fe coverage and because the anisotropy energy drops rapidly when the Curie temperature is approached, we scaled the experimental values by a factor of 10 in most cases to have the same order of magnitude for both panels of Fig. 5.8. Apart from the temperature there are other possible reasons for the smaller experimental MAE that are partly hard to quantify or to account for in our calculations. First of all, one has to mention the detailed structure at the interface due to alloying and reconstruction. Another possible reason for the smaller experimental MAE – the influence of protective cap layers (in general Au) on top of the Fe film – will be investigated in Sec. 5.5.2.

5.4.4   In-plane magnetic anisotropy

Comparing the orientations of the magnetization along the [110]- and [1$\bar{1}$0]-directions in the plane there is no shape anisotropy contribution, because both directions are equivalent if the dipole-dipole interaction is dealt within a classical way. As a consequence only the electronic part of the MAE has to be considered. As the left panel of Fig. 5.9 shows, apart from n=2, the [110]-direction of the magnetization is clearly favored by the electronic part of the MAE. The experimental situation, on the other hand, is less clear cut. In several investigations [192, 201–204] the [1$\bar{1}$0]-direction was reported to be the easy axis for thin Fe films on GaAs. The majority of experiments [191, 193, 195–200], however, found the [110]-direction to be the easy axis. The corresponding experimental MAE data have been collected in the right panel of Fig. 5.9. As most of these data have been obtained by room temperature measurements, we again have scaled them up by a factor of 10 (see discussion above). This

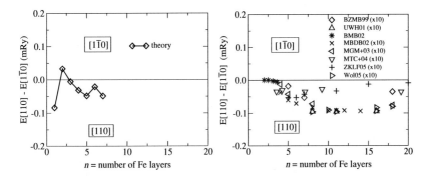

Fig. 5.9. MAE for GaAs/nFe systems with n=1–7, given as the difference between an orientation along the [110]- and [1$\bar{1}$0]-directions. The left panel shows theoretical results, while the right panel shows experimental data partly scaled by a factor 10 (see text). References to the experiments: BZMB99 [193], UWH01 [197], BMB02 [196], MBDB02 [198], MGM+03 [201], MTC+04 [199], ZKLF05 [200] and Wol05 [227].

obviously leads to values that fit reasonably well to the results of Bensch *et al.* [196], that have been obtained for 10K. It was suggested that the discrepancy among the experimental data concerning the orientation of the easy axis has to be ascribed to a different reconstruction of the GaAs surface prior to the deposition of Fe. This could be ruled out by the work of Kneedler *et al.* [191] who prepared Fe films on As-terminated 2×4- and c(4×4)-reconstructed surfaces and of Moosbühler *et al.* [198] who prepared Fe films on Ga-terminated 2×6- and 4×2-reconstructed surfaces. In both investigations the [110]-direction was found to be the easy axis, independent on the termination and on the reconstruction. In addition, the Curie temperatures for Fe films on Ga-terminated 2×6- and 4×2-reconstructed surfaces were found to be the same [205]. From these findings it can be concluded that the deposition of Fe on a GaAs leads to an interface structure that hardly depends on the previous surface reconstruction of the substrate. Furthermore it justifies our structural model to some extent which assumes a perfect GaAs–Fe interface. In addition it should be noted that test calculations for few As-terminated substrates led essentially to the same results as for the Ga-terminated systems.

Our calculations based on the adopted geometry model not only give the correct orientation of the easy axis but also show that the interface contribution to the MAE is most important. This can be demonstrated by a layer by layer decomposition of the MAE. The results for GaAs/7Fe system shown in Fig. 5.10 indeed show that the dominating contributions to the MAE stem from the Fe layers close to the interface. In addition there is an appreciable contribution from the Fe surface layer. This behavior is very similar to that found before for SC/FM multilayer systems [78–80] and also for metallic bilayer systems [77]. In line with our results Bensch *et al.* [196] finds in experiment that the uniaxial anisotropy is most pronounced for thin Fe films ($n < 15$) while a change of the anisotropy according to the fourfold symmetry of thick Fe films is found when the film thickness is increased beyond $n \approx 30$ [195].

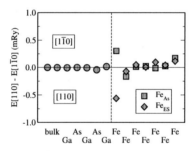

Fig. 5.10. Layer-resolved MAE for the GaAs/7Fe system, given as the difference between an orientation along the [110]- and [1$\bar{1}$0]-directions. The negligible contributions of the empty spheres and vacuum region have been omitted.

While interface alloying [208, 211, 212, 217] and strain [201, 204] will certainly influence the anisotropy, the consistent picture that emerges from our results and the experimental work of Kneedler *et al.* [191] and Moosbühler *et al.* [198] clearly demonstrates that these effects are of minor importance for Fe films on GaAs substrate. Another point, the influence of a covering layer on the MAE of GaAs/nFe systems will be investigated in the following Sec. 5.5.

5.5 Influence of Au covering layers

5.5.1 Introduction

There is a large number of experimental investigations that dealt with the magnetic anisotropy of layered SC/FM systems, with a strong emphasis on Fe films on (001) oriented GaAs [191–204]. In most experiments listed above, the surface of Fe on GaAs was covered by protective layers of a noble metal (in most cases Au) to avoid the corrosion of Fe. Independent on the type of the covering layer, all experiments show a strong uniaxial (twofold) in-plane anisotropy with the easy axis in the [110]-direction and the hard axis in the [1$\bar{1}$0]-direction.

In the experimental work of Bayreuther et $al.$ [88] it was found that the biaxial (fourfold) in-plane anisotropy is not affected by the Au overlayer. Moreover, it was shown that there is no uniaxial in-plane contribution to the MAE in Fe films on Au substrate [87]. Therefore it was deduced that the Au covering layer has no impact on the uniaxial MAE in the GaAs/Fe/Au system [193]. On the other hand, in the in $situ$ Brillouin Light Scattering (BLS) experiment of Madami et $al.$ [199] it could be shown that the presence of a Cu covering layer causes an appreciable reduction of the in-plane uniaxial anisotropy. In the following, we investigate the influence of the Au overlayer calculating the MAE and assign it to the atoms in the GaAs/Fe/Au systems.

Including Au overlayers in calculations allows to examine the effect of the Au covering layer on the uniaxial MAE. There have been two types of Au covering layers investigated – the fcc-like and bcc-like ones. One can assume that thick Au covering layers, used in most experiments, order in the fcc structure and for that reason 3 fcc Au layers have been included on top of the Fe film on GaAs in the calculation (GaAs/nFe/3 fcc Au systems with Fe thickness n=1–7) in Sec. 5.5.2. In this section we will concentrate on the influence of fcc Au covering layers on top of Fe films on GaAs by comparing the out-of-plane and in-plane MAE of uncovered GaAs/Fe systems (previous Sec. 5.4) and covered GaAs/Fe/fcc Au systems. In addition, the layer by layer decomposition of the MAE will give a clearer view on the issue.

The experimental team in Perugia University has examined the effect of a thin Au covering layer on the anisotropy of GaAs/Fe films by exploiting the *in situ* BLS. Assuming that the first Au layers growing on top of bcc Fe adopt the bcc structure, bcc Au layers were used to model ultrathin Au overlayers (GaAs/4(7)Fe/bcc Au systems). A semi-quantitative interpretation of the experimental investigations on ultrathin bcc-like Au overlayers is given by the TB-KKR method of band structure calculation in Sec. 5.5.3.

It will be shown that the fcc-like Au covering layers do not influence the uniaxial in-plane anisotropy significantly, whereas the bcc-like Au overlayers strongly reduce the uniaxial MAE. This is to our knowledge the first theoretical study which includes covering layers on SC/FM systems.

5.5.2 GaAs/nFe/3 fcc Au systems, n=1–7

5.5.2.1 Out-of-plane magnetic anisotropy

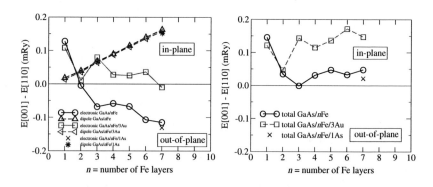

Fig. 5.11. MAE for GaAs/nFe and GaAs/nFe/3Au systems with n=1–7, given by the difference between out-of- and in-plane orientation of magnetization. The left panel shows the electronic and dipole-dipole part of the MAE. The right panel shows the sum of the two theoretical contributions to the MAE.

The results for the MAE between the out-of- and in-plane magnetization direction obtained for the GaAs/nFe and GaAs/nFe/3Au systems with n=1–7 are shown in Fig. 5.11. As one can see, the electronic part of MAE favors an out-of-plane orientation for the GaAs/nFe system – except for the case of Fe monolayer (n=1) – and an in-plane orientation for the GaAs/nFe/3Au system. To compare the results for the MAE with corresponding experimental data the shape anisotropy (dipole-dipole) contribution has to be added to the electronic part. Comparing the out-of- and in-plane orientation of $\vec{M}$ one finds the shape anisotropy to vary nearly linearly with the number of Fe layers for the investigated range of n, being similar for the GaAs/nFe and GaAs/nFe/3Au system (see left panel of Fig. 5.11). The sum of the electronic and dipole-dipole part is plotted in the right panel of Fig. 5.11. The in-plane magnetization direction has been observed in the experiments (see Fig. 5.8) and could be obviously reproduced in our theoretical study.

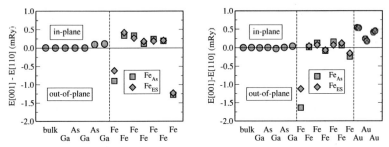

Fig. 5.12. Layer-resolved electronic MAE for the GaAs/7Fe and GaAs/7Fe/3Au systems, given by the difference between out-of- and in-plane orientation of magnetization. The negligible contributions of the empty spheres and vacuum region have been omitted.

The covering of the Fe surface by Au layers leads to a higher energy difference between the out-of- and in-plane magnetization direction in the GaAs/Fe/3Au systems compared to the uncovered Fe films on GaAs (GaAs/Fe systems). The layer-resolved electronic MAE values of the GaAs/7Fe system in Fig. 5.12 show that mostly the Fe interface and surface layers contribute to the out-of-plane direction of $\vec{M}$. On the other hand, in the GaAs/7Fe/3Au system the Fe layer at the GaAs–Fe interface contributes, whereas the Fe layer at the Fe–Au interface does not contribute significantly. The coverage by Au results in the reduction of the electronic out-of-plane tendency in the GaAs/Fe/3Au systems. It is in accordance with the experimental finding of Purcell *et al.* [228] that the perpendicular anisotropy constant with the preferable out-of-plane magnetization direction has weakened after deposition of Au on a thick Fe film.

5.5.2.2 In-plane magnetic anisotropy

As discussed already, the preferable direction of the magnetization in thin Fe films on GaAs substrate is in the plane, independent of the coverage. Comparing two perpendicular directions in the plane there is no shape anisotropy contribution, because both directions are equivalent if the dipole-dipole interaction is dealt in a classical way. As a consequence only the electronic part of the MAE has to be considered. As Fig. 5.13 shows, apart from $n=2$ for the GaAs/nFe system, the [110]-direction of the magnetization is clearly favored by the electronic part of the MAE. After comparing the in-plane MAE for GaAs/nFe and GaAs/nFe/3Au systems one can see that the in-plane behavior remains essentially unchanged after the coverage by 3 layers of fcc Au.

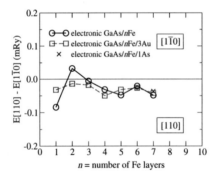

Fig. 5.13. MAE for GaAs/nFe and GaAs/nFe/3Au systems with $n=1$–7, given as the difference between an orientation along the [110]- and [1$\bar{1}$0]-directions.

It was deduced that a coverage by thick Au overlayer should not influence the in-plane MAE in the GaAs/Fe system [193]. This assumption was supported by the observation of Krebs $et\ al.$ [202] and Sjöstedt $et\ al.$ [80] that the biggest contribution to the in-plane MAE is stemming from the anisotropic interface bonding between GaAs and Fe. To support these observations, the layer-resolved in-plane MAE of

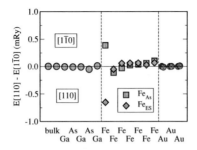

Fig. 5.14. Layer-resolved electronic MAE for the GaAs/7Fe/3Au system, given by the difference between an orientation along the [110]- and [1$\bar{1}$0]-directions. The negligible contributions of the empty spheres and vacuum region have been omitted.

the GaAs/7Fe/3Au system is presented in Fig. 5.14. It shows that the Fe contribution at the GaAs–Fe interface is most important. The contribution from the Fe–Au interface of the GaAs/7Fe/3Au system and from the Fe surface in GaAs/7Fe system (see Fig. 5.10) are similar. As the contribution from the 3 fcc Au layers is negligible, no change of the in-plane MAE after coverage by fcc Au is expected.

In several experiments [198, 212, 229] diffusion of As atoms through the Fe layers has been observed. The As atoms segregate at the surface and form an overlayer. This behavior has been theoretically studied by Mirbt *et al.* [230]. In order to model and explore the influence of an As overlayer in addition, a monolayer of As on top of 7 Fe layers on GaAs (GaAs/7Fe/1As system) has been included. As shown in Fig. 5.11, the As overlayer causes a moderate decrease of the MAE between the out-of-plane and in-plane magnetization direction. Further, in Fig. 5.13 no change of the preferable magnetization direction and no substantial effect on the energy differences between the [110]- and [1$\bar{1}$0]-directions has been observed.

5.5.3   GaAs/4(7)Fe/bcc Au systems

5.5.3.1   Brillouin light scattering (BLS)

The mostly used experimental techniques for the *in situ* magnetic characterization are the Brillouin light scattering (BLS) and the surface magneto-optical Kerr effect (SMOKE) magnetometry. The first technique is a dynamic probe, testing the propagation properties of long-wavelength light. Using the second technique, on the other hand, the static response of a magnetic system is studied, by an inspection of the hysteresis loop.

In the last decade Brillouin light scattering has proved to be a powerful technique for the study of spin waves in two-dimensional systems, with a sensitivity to the monolayer scale [231–234]. Since the magnetic properties are strongly structure-sensitive, the possibility of performing *in situ* Brillouin light scattering is especially attractive, for understanding the interrelation between the structural properties and the magnetic behavior of a thin film. Indeed, thanks to ultra-high-vacuum (UHV) conditions, which prevent sample contamination, *in situ* measurements allow the direct study of uncovered films, avoiding any possible structural changes which can be caused by either exposure to air or a protective capping layer. Apart from the pioneering BLS measurements of Hillebrands *et al.* [235, 236] on Fe/W(110) and of Scheurer *et al.* [237] on Fe/Cu(100), however the only recent combined *in situ* BLS and SMOKE experiments dealt with Fe/Ag(100) films [238, 239]. This lack of data can be due to the fact that the BLS alignment and measurement procedure may become complex and time-consuming in too large UHV systems.

BLS usually refers to the inelastic scattering of light from thermally activated acoustic phonons, as well as spin waves. In a BLS experiment, a beam of monochromatic light is focused on the sample under investigation, and the light scattered within a solid angle is frequency-analyzed. Since the wavelength of the revealed spin wave is of the same order of magnitude as that of light, the frequency of the spin wave, probed in a typical Brillouin light scattering experiment, is in the range between 1 and 50 GHz. Therefore, a high-resolution spectrometer, like a multi-pass Fabry-Perot interferometer, is necessary in order to extract the weak inelastic component of light from the elastically scattered contribution. In magnetic materials the frequency shift of scattered light is caused by the interaction between the electric field of the incident optical wave, and the sample with its refractive index modulated by the precessing magnetization. According to this interaction, a distribution of electric dipoles oscillating at a frequency equal to the optical one, plus and minus that of the magnetic normal modes, is generated. Then, by means of a Brillouin light scattering experiment, the frequency of spin wave can be directly determined. It is important to point out that the light does not excite the magnetic mode, but it is exploited only as a probe of the spin wave, which is already present inside the film. From Brillouin measurements of the spin wave frequencies as a function of the direction and magnitude of the in-plane wave vector, and of the direction and strength of the externally applied field, a determination of magnetic parameters, such as saturation magnetization, anisotropy constants as well as film thickness, can be achieved. Specific advantage of BLS is that it is a non-destructive technique, characterized by high sensitivity. Moreover, thanks to the small laser focus size, BLS is a local probe which can be used to map the sample surface and check lateral variation of parameters like film thickness.

5.5.3.2 Experimental setup

The sample preparation and all measurements were performed at the Group of High Resolution Optical Spectroscopy and related Techniques (GHOST) of Perugia University (the group of Prof. G. Carlotti). An ultra-high-vacuum (UHV) chamber allowed *in situ* BLS measurements [240]. Four Fe films of different thickness (n=4, 7, 20 and 33ML) were evaporated on GaAs(001) single crystal wafers. After the deposition of Fe the crystallographic quality of Fe films was checked by Low-Energy Electron Diffraction (LEED). For all samples a sharp LEED pattern has been observed ensuring a well ordered bcc (001) structure. Further details about structural characterization of Fe film grown on the GaAs(001) substrate can be found elsewhere [199]. As the main aim was to study the influence of a nonmagnetic Au covering layer on the in-plane anisotropy, 2ML of Au have been evaporated on the Fe films in addition.

BLS measurements were performed *in situ* by positioning the specimen close to an optical viewport. The back-scattered light was analyzed by a multi-pass Fabry-Perot interferometer. The external d.c. magnetic field was applied parallel to the film surface and perpendicular to the plane of incidence of light.

5.5.3.3 Experimental results

The BLS measurements have been first carried out for the uncovered films. A typical BLS spectrum relative to the 20ML thick Fe film, is shown as an inset of Fig. 5.15c. In addition to the dominant peak, due to the elastically scattered light, the so called Damon-Eshbach (DE) spin-wave mode [241] is clearly seen on both sides of the spectrum. Due to the relatively low values of the film thickness, only this mode is present in the spectra, characterized by a remarkable Stokes–anti-Stokes intensity asymmetry, typical of magnons in thin ferromagnetic films of absorptive materials [242]. The dependence of spin-wave frequency on the in-plane direction of the applied magnetic field with respect to the [100]-direction of the GaAs(001) substrate (ϕ_H) is shown in Fig. 5.15 for the four different films. In agreement with previous studies, a strong in-plane uniaxial anisotropy, with the easy axis parallel to the [110]-direction, has been observed in the thinnest samples (GaAs/nFe with n=4 and 7). On increasing the thickness to n=20 the expected cubic anisotropy of bulk Fe starts to develop, until it becomes the predominant contribution for n=33.

To study the influence of a nonmagnetic Au covering layer on the in-plane anisotropy, 2ML of Au have been evaporated on the Fe films and the spin-wave frequency has been measured again as a function of the in-plane direction of the magnetic field. Fig. 5.15 shows that the presence of the Au overlayer, although only 2ML thick, affects the system substantially. In particular, no magnetic signal could be measured from the 4ML thick film, indicating that ferromagnetism is destroyed by the Au overlayer at room temperature. Fig. 5.15b shows a spin-wave frequency independent of the in-plane angle for the Au overlayer. The 7ML thick Fe film is ferromagnetic after the Au deposition, but a total suppression of the in-plane uniaxial anisotropy was observed. On increasing the Fe thickness the suppression effect becomes less ef-

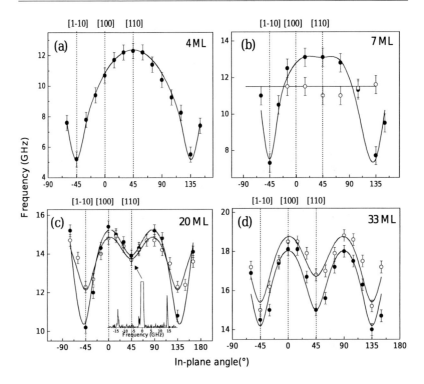

Fig. 5.15. Experimental spin-wave frequencies as a function of the in-plane angle ϕ_H for uncovered (filled circles) and covered (open circles) films of different thickness. Continuous curves denote the calculated spin-wave frequencies obtained by fitting the BLS data. All measurements were performed applying an external field of 1.0 kOe and with an incidence angle of light of $20°$. A Brillouin spectrum, taken along the [110] in-plane direction, is shown for the 20ML thick Fe film in panel c.

fective, although it is surprisingly still appreciable in the thicker films. Moreover, concerning the influence of the Au thickness on the in-plane anisotropy, the same results were obtained on increasing the Au thickness from 2ML to 20ML.

To quantify the effect of the Au capping layer in a better way, a best fit procedure of the experimental data to the calculated frequencies was performed, using the analytical expression of the DE mode frequency

n (ML)	K_2 (10^5 erg/cm^3)	K_4 (10^5 erg/cm^3)
4	5.0±0.6	-0.16±0.01
(Au covered)	—	—
7	3.5±0.3	1.2±0.3
(Au covered)	—	—
20	2.2±0.2	2.0±0.2
(Au covered)	1.1±0.2	1.3±0.2
33	0.5±0.2	2.5±0.2
(Au covered)	0.8±0.2	1.8±0.2

Table 5.1. Values of the nominal thickness n of Fe films, of the effective magnetization ($4\pi M_{\text{eff}}$) and of the in-plane uniaxial (K_2) and biaxial (K_4) phenomenological anisotropy constants.

in the ultrathin film approximation [232, 243]. This procedure enabled the determination of both the phenomenological in-plane uniaxial (K_2) and in-plane biaxial (K_4) anisotropy constants, whose values are listed in Table 5.1. The obtained values of the anisotropy constants confirm that the Au overlayer strongly influences the uniaxial anisotropy in the Fe films with thickness up to 20ML, while the effect becomes less important in the 33ML thick film. Moreover, the best fit procedure points out that the Au covering layer leads to a partial suppression of the biaxial anisotropy contribution too. In such a case, however, the degree of suppression remains almost constant on increasing the Fe thickness.

5.5.3.4   Theoretical investigation

To achieve an interpretation of the main experimental result, i.e. the above mentioned reduction of the uniaxial in-plane anisotropy caused by the thin Au overlayer, the magnetic anisotropy energy for the two thinnest samples (GaAs/4(7)Fe/m bcc Au system) has been calculated using the fully-relativistic version of the tight-binding-KKR (TB-KKR) method of band structure calculation (see Sec. 2.3). As will be shown in the following, with this method the decrease of the uniaxial anisotropy in the GaAs/Fe films covered with bcc Au in qualitative agreement with the experimental data could be verified, while the loss of magnetization in the GaAs/4Fe/Au system could not be explained. However, one has to keep in mind that the calculations are done for zero temperature for an ideal structure, with no interdiffusion, no roughness and no interlayer distance relaxations taken into account. This prevents from a direct quantitative comparison between the calculated and the measured anisotropy constants, which are in fact almost an order of magnitude smaller than the calculated ones.

5.5.3.5 Magnetic anisotropy energy and Bruno's expression

The electronic MAE is determined as the difference in total energy for two different orientations of the magnetization $\vec{M}$ in the system. As the in-plane magnetic anisotropy is of interest here, two magnetization directions in the film plane will be compared: the [110]-direction (easy axis) and the [1$\bar{1}$0]-direction (hard axis).

Among the various approaches used to resolve the MAE into its atomic contributions Bruno's formula according to Eq. (2.98) will be used in the following. In this equation ξ is the spin-orbit coupling parameter and the difference in the brackets of Eq. (2.98) is nothing else but the orbital moment anisotropy of the atoms under investigation. The sum over all atoms will give the total MAE. For an application of this expression, for all Fe, Ga, As and Au atoms the values of the spin-orbit coupling parameters ξ of the corresponding bulk systems (ξ_{Ga-p}=15 mRy, ξ_{As-p}=22 mRy, ξ_{Fe-d}=4.8 mRy and ξ_{Au-d}=71 mRy) have been used.

5.5.3.6 In-plane magnetic anisotropy

The solid lines in Fig. 5.16 show the difference in total energy for the [110] easy axis and the [1$\bar{1}$0] hard axis for both the GaAs/4Fe/mAu and the GaAs/7Fe/mAu systems. In case of the GaAs/7Fe/mAu systems the uniaxial anisotropy decreases nearly linearly with the Au coverage, whereas for GaAs/4Fe/mAu it oscillates strongly. In spite of this oscillation, however, an overall suppression of the uniaxial anisotropy can be observed on increasing the Au thickness. The atomic layers responsible for the oscillatory behavior can be identified by using Eq. (2.98) for resolving the magnetic anisotropy energy (MAE) into its atomic contri-

butions. First, the validity of Eq. (2.98) has been tested by comparing
the MAE from the total energy difference (solid lines in Fig. 5.16) and
the MAE obtained from Bruno's formula using Eq. (2.98) (dashed lines
in Fig. 5.16). As Bruno's formula is an approximation, it reproduces
the proper total energy results only in a qualitative way. Nevertheless,
it is able to give the linear behavior in the GaAs/7Fe/mAu systems
and the oscillatory behavior in the GaAs/4Fe/mAu systems.

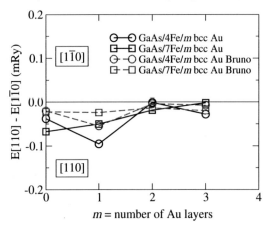

Fig. 5.16. The dependence of the in-plane magnetic uniaxial anisotropy on the number of
Au covering layers ($m = 0$ stands for an uncovered Fe film on GaAs). The solid lines
show the difference in total energy for two different orientations of the magnetization $\vec{M}$.
The dashed lines, on the other hand, are connected to the anisotropy of the orbital angular
momentum according to Eq. (2.98).

After the successful test of Bruno's formula, the MAE of the
GaAs/4Fe/mAu and GaAs/7Fe/mAu systems could be resolved into its
atomic contributions as given in Fig. 5.17. The contributions from the
GaAs (squares) and Fe layers (diamonds) are more or less constant for
the whole sequence of Au coverages. The contribution from the GaAs
layers is the smallest one because of the low orbital moment anisotropy
of Ga and As atoms at the GaAs–Fe interface. As the induced orbital
moment anisotropy in the Au layer at the Fe–Au interface is of the
same magnitude as the orbital moment anisotropy in the Fe layers, and
the spin-orbit coupling of Au atoms is 15 times bigger than that of the

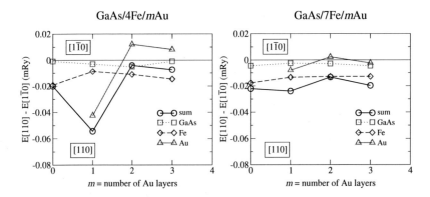

Fig. 5.17. Atom-resolved MAE of the GaAs/4Fe/mAu (left panel) and GaAs/7Fe/mAu systems (right panel) obtained by using Eq. (2.98). The sum over all atomic contributions (spheres) corresponds to the dashed lines in Fig. 5.16.

Fe atoms, there is a considerable contribution from the Au layers to the total MAE of the GaAs/Fe/Au systems (see triangles in Fig. 5.17). In the GaAs/4Fe/mAu system (left panel of Fig. 5.17), the Au contribution is responsible for the oscillation of the total MAE. In the case of the GaAs/7Fe/mAu system (right panel of Fig. 5.17) the Au contribution is much smaller than in the GaAs/4Fe/mAu system and it hardly influences the total uniaxial anisotropy, as the uniaxial anisotropy is a GaAs–Fe interface effect. For the thicker Fe film (GaAs/7Fe/mAu system) the Au overlayers are further away from the GaAs–Fe interface and therefore their contribution is smaller in magnitude than for Au overlayers deposited on the thinner Fe film (GaAs/4Fe/mAu system).

5.6   Co marker in Fe film on GaAs

5.6.1   Introduction

In order to examine the layer-resolved magnetic properties of the Fe film on GaAs substrate in an experiment, Giovanelli *et al.* [6–8] introduced a Co marker in the Fe film. By seeding half a monolayer of Co in a 7 ML thick Fe film on GaAs, a monitoring of the magnetic properties of the Fe film as a function of the distance from the GaAs–Fe interface was possible. Already 20 years ago, layer dependent analysis of magnetic films has been demonstrated by selectively depositing isotopes of Fe and Co during the magnetic film growth and performing subsequently Mößbauer spectroscopy experiments [244]. In the work of Giovanelli *et al.* [6–8] an alternative approach to study the magnetic properties at, and in the neighborhood of, the GaAs/Fe interface was reported. X-ray magnetic circular dichroism (XMCD) gives access, via the chemical sensitivity of X-ray absorption, to both spin and orbital magnetism, and can be sensitive to small concentrations of impurities deposited at various distances from the relevant interface. Previous studies suggest that Co is the proper magnetic marker in Fe on GaAs because: i) $Fe_{0.50}Co_{0.50}$ on GaAs has an identical lattice parameter as Fe on GaAs [245] and ii) the magnetic properties of the thin film are unperturbed by the presence of Co [195]. Moreover, Co impurities in bcc Fe as well as diluted Co-Fe bcc alloys show that the Co magnetic moment is strongly influenced by the magnetism of the surrounding Fe [246].

The measured evolution of spin and orbital moment across the ferromagnetic film shows an enhancement of the orbital moment near the interface with GaAs, as well as a general decrease of the spin magnetization when approaching the interface and the film surface. The theoretical investigation of the spin and orbital magnetization of the Co marker in a Fe film on GaAs and comparison with the experimental results will be the topic of this section. First, we will describe the experimental double wedge with half-monolayer Co marker along the Fe film and introduce the corresponding theoretical model with various depth of the $Fe_{0.50}Co_{0.50}$ layer in the ferromagnetic film. In the following, the XMCD sum rules allowing the connection of the XMCD spectra with the spin and orbital moments will be discussed, mainly the experimentally unknown parameters – the number of unoccupied d–states N_{hd} and the spin dipole moment, the so-called T_z term. After delivering our calculated N_{hd} to the experimental values, agreement between the experimental and theoretical spin moments could be observed. The aim of the experimental study was to detect the magnetization profile of the Fe film with the help of the Co atomic marker. The validity of this assumption will be discussed at the end.

5.6.2 Studied systems

Experiments were performed at the APE-INFM beamline at the ELETTRA synchrotron radiation laboratory in Trieste. A double wedge structure of Fe containing an oblique layer of $Fe_{0.50}Co_{0.50}$ was prepared as follows. A moveable shutter allowed to grow a first Fe wedge of 6 atomic layers thickness over a sample. On this wedge a single monolayer of $Fe_{0.50}Co_{0.50}$ was deposited by co-evaporation. Finally an opposite wedge of pure Fe was grown obtaining a sample with Co atoms embedded at a variable distance (denoted as n) from the interface in an evenly thick 7 ML film, as sketched in the upper panel of Fig. 5.18. This way, the Co magnetic markers were embedded in the macroscopically magnetized Fe film, and their XMCD signal reflected the local magnetization of the Fe film that changes from the interface up to the surface with a characteristic profile. The samples were prepared *in situ* under UHV conditions, thus avoiding the uncertainties introduced by covering layers.

In order to interpret the experimentally achieved spin and orbital moments, the fully-relativistic version of the tight-binding Korringa-Kohn-Rostoker (TB-KKR) method of band structure calculation (see Sec. 2.3) has been used. The experimental setup, e.g. the double wedge with the half monolayer of Co at variable distance from the GaAs–Fe interface and the total thickness of 7ML for the ferromagnetic film has been modelled as follows. In the lower panel of Fig. 5.18 the 7 theoretically investigated systems with different positions of the $Fe_{0.50}Co_{0.50}$ layer in the ferromagnetic film are sketched. One can denote these systems in general as $GaAs/(n-1)Fe\,(Fe_{0.50}Co_{0.50})\,(7-n)Fe$ with $n=1$–7, with n being the position of the Co marker in the Fe film. $n=1$

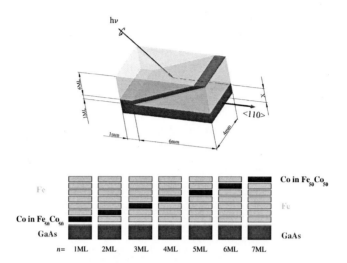

Fig. 5.18. Upper panel: Sketch of the experimental geometry of the double wedge. Co atoms were embedded at a variable distance (n) from the interface in an evenly thick 7 ML Fe film. The sample was magnetized in-plane along the easy axis, the [110]-direction. Lower panel: The 7 theoretically investigated systems with different positions of the $Fe_{0.50}Co_{0.50}$ layer in the ferromagnetic film.

stands for the $Fe_{0.50}Co_{0.50}$ layer at the GaAs interface and $n=7$ for the $Fe_{0.50}Co_{0.50}$ layer at the film surface.

As reported in several experiments [198, 212, 229] and one theoretical study [230], an As segregation towards the Fe surface takes place even at 0K. Therefore the impact of an As overlayer has been studied in addition. This was achieved by modelling a half-monolayer of As on top of the ferromagnetic film. As we intend to study effect of the As overlayer on the Co marker, the As overlayer was included only for the systems with $Fe_{0.50}Co_{0.50}$ layer near the film surface ($n=6$ and 7).

5.6.3   XMCD sum rules

The X-ray magnetic circular dichroism (XMCD) that is caused by spin-orbit coupling is one of the most powerful experimental tools to investigate the magnetic properties of complex systems in an element specific way. This is in particular a result of the successful application of the XMCD sum rules, derived and discussed by several authors [134, 221, 223, 247].

On the basis of Eq. (5.1) for the X-ray absorption coefficient it was possible to link the integrated absorption spectra for a specific core shell and polarization of the radiation to the expectation values of the operators σ_z and l_z and in that way to the spin and orbital moments, μ_{spin} and μ_{orb}, of the absorber atom. In spite of the very appealing features of the XMCD sum rules, one has to keep in mind that in deriving these rules a large number of assumptions had to be made [223]. For the $L_{2,3}$-edge these expressions – the XMCD sum rules – have been derived by Carra $et\ al.$ [134] and Thole $et\ al.$ [221]:

$$\int \left(\Delta\mu_{L_3} - 2\Delta\mu_{L_2} \right) dE = \frac{N}{3N_{hd}} \left(\langle \sigma_z \rangle_d + 7 \langle T_z \rangle_d \right) \qquad (5.2)$$

$$\int \left(\Delta\mu_{L_3} + \Delta\mu_{L_2} \right) dE = \frac{N}{2N_{hd}} \langle l_z \rangle_d \ , \qquad (5.3)$$

where N is the integrated spectrum for unpolarized radiation. Eqs. (5.2) and (5.3) contain two experimentally unknown parameters – the N_{hd} and the so-called T_z term. N_{hd} is the number of unoccupied d–states, i.e., of d–holes, and $\langle T_z \rangle$ is the expectation value of the magnetic dipole operator, which can be seen as a measure for the asphericity of the spin magnetization. While the T_z term is often negligible for bulk systems with high symmetry, this does not apply to plain surfaces and cluster atoms at the surface. In the work of Minár $et\ al.$ [248] the influence of the parameters N_{hd} and T_z to cluster atoms was investigated. The N_{hd} of atoms in a Co_n cluster only slightly oscillated around the bulk value of hcp Co. This showed that the value of N_{hd} deduced from calculations on a reasonable reference material can be seen as a good estimate for surface systems. A more pronounced variation of the so-called T_z term was expected for atoms in a Co_n cluster. The calculations gave $-0.15\mu_B$ for the Co-adatom and about $-0.01\mu_B$ for the 5-atom cluster. The later value was already quite close to the results for bulk hcp Co.

One has to note, that the T_z term in Eq. (5.2) contributes weighted by a factor of 7 to the sum rule analysis. Thus, even a small value of the

T_z term can influence the estimate for the spin magnetic moment μ_{spin} in a substantial way. The results indeed show that because of the low symmetry of the surface the T_z term can in general no longer be ignored in an application of the sum rules to deduce the spin magnetic moment from experimental XMCD spectra. The orbital moment μ_{orb} and the T_z term depend on the direction of the magnetization. Therefore, the magnetization direction in the calculations was chosen to correspond to the experimental setup. It was parallel to the easy axis, the [110]-direction.

5.6.4   Spin and orbital magnetization profiles

In the left upper panel of Fig. 5.19 the spin moment (μ_{spin}) of the Co marker at various distance from the interface is shown. The first three data sets show the raw experimental data, then the influence of the calculated N_{hd} value and of the T_z term. In the estimation for μ_{spin} of the Co marker in the XMCD experiment by Eq. (5.2) the experimental estimate for N_{hd} of 1.62 was used. This estimate was determined by measuring the coherent scattering of a polarized neutron beam by Di Fabrizio *et al.* [249] as follows. From derived magnetic moments at the Fe and Co sites the charge distribution of d electrons in different subbands was detected, after considering that the majority $3d$ bands are fully occupied. However, as shown already 20 years ago by Schwarz *et al.* [250], the majority $3d$ band of Co in FeCo alloy is not completely occupied (the number of majority $3d$ electrons is 4.66 instead of 5.0). In the following, using our calculated N_{hd} value of 2.45 instead of the estimate 1.62, the experimental μ_{spin} and μ_{orb} values increased by a factor of ca. 3/2 (see the solid line "exp - theor N_{hd}" in Fig. 5.19). This brings the experimental μ_{spin} close to the theoretical results. Due to the high symmetry in the center of the ferromagnetic film, the T_z term is expected to have minor importance. On the other hand, at the film surface the T_z term has a value of about 0.05 μ_B and it leads to an additional decrease of the experimental μ_{spin} at the surface (see Fig. 5.19).

In the experiment the reduction of μ_{spin} of the Co marker at the GaAs–Fe interface was attributed to the Ga and As intermixing with the ferromagnet. A decrease of the μ_{spin} could be observed even without

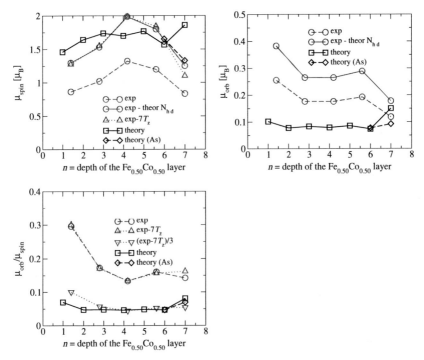

Fig. 5.19. Comparison of the theoretical and experimental results. Upper panel: Spin and orbital moments of the Co marker atoms depending on the distance from the GaAs–Fe interface. Lower panel: The orbital-to-spin moment ratio. The legend follows as: "exp" – raw experimental results, "exp - theor N_{hd}" – experimental results after corrected number of holes, "exp - $7T_z$" – the T_z term taken into account, "theory" – theoretical results, "theory (As)" – theoretical results including the As overlayer.

interdiffusion of the Ga and As atoms in our theoretical study. Surface segregation and accumulation of As on top of the Fe film was reported in previous experiments [198, 212, 229]. After including the As overlayer in calculations, the reduction of μ_{spin} at the film surface has been found. In summary, we showed that the main features of μ_{spin} of the Co marker – the decrease at the interface and surface – could also be demonstrated theoretically.

As shown in the right panel of Fig. 5.19 the enhancement of μ_{orb} at the interface could be observed by experiment and theory. The decrease of μ_{orb} at the surface could not be detected in the calculations. The increase of the calculated μ_{orb} could be at least damped after including the As overlayer. As already mentioned in Sec. 3.2.3, the plain LSDA scheme neglects some many body effects which give rise to the Hund's second rule in atomic theory. As the calculated μ_{orb} is much too low in comparison with the experiment, we could compare the calculated and experimental values only qualitatively.

From the Eqs. (5.2) and (5.3) follows that the $\mu_{\mathrm{orb}}/\mu_{\mathrm{spin}}$ ratio is independent of the number of holes N_{hd}. That makes the problems connected with the experimentally unknown N_{hd} value not relevant. Because of the higher experimental μ_{orb}, also the experimental $\mu_{\mathrm{orb}}/\mu_{\mathrm{spin}}$ ratio is much higher than the theoretical one. Therefore the experimental data in the lower panel of Fig. 5.19 have been divided by an arbitrary factor of 3 enabling a qualitative comparison. Also the small impact of the T_z term can be seen there. The main features of the $\mu_{\mathrm{orb}}/\mu_{\mathrm{spin}}$ ratio, observed by experiment and calculations, is the enhancement at the interface and a constant behavior through the center of the film up to the surface. Finally, the effect of the As overlayer is canceled in the $\mu_{\mathrm{orb}}/\mu_{\mathrm{spin}}$ ratio.

5.6.5   Co marker detecting the Fe magnetization profile

In the experiment it was suggested that the Co marker can monitor the magnetic properties of the surrounding Fe atoms and therefore it is suitable for detecting the Fe magnetization profile in a Fe film on GaAs. In the following we will investigate how suitable the Co marker is. In Fig. 5.20 the μ_{spin} and μ_{orb} of the Co marker in $\mathrm{GaAs}/(n-1)\mathrm{Fe}\,(\mathrm{Fe}_{0.50}\mathrm{Co}_{0.50})\,(7-n)\mathrm{Fe}$ system and of the Fe atoms in GaAs/7Fe system are shown in parallel. The magnetization profile of the GaAs/7Fe system has been already sufficiently discussed in Sec. 5.4.1. The decrease of μ_{spin} at the interface in the case of the Co marker can not be observed in the GaAs/7Fe system. The increase at the surface of the GaAs/7Fe system is more pronounced and including the As overlayer leads only to a constant value of μ_{spin}. On the other hand, the behavior of μ_{orb} is well reproduced by the Co marker.

As shown, the Co marker describes the behavior of the inner layers of the Fe film correctly. The strong increase of μ_{orb} at the GaAs–Fe

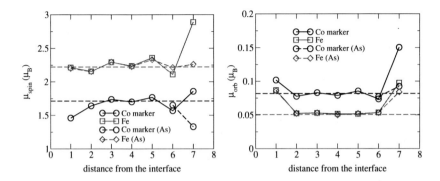

Fig. 5.20. Theoretical results manifesting the reliability of the Co marker in detecting the Fe magnetization profile of the GaAs/7Fe system. The horizontal dashed lines stand for the bulk Fe and Co values. (As) in the legend means the systems with an As overlayer.

interface and surface of the film is well reproduced by the Co marker. On the other hand, the behavior of μ_{spin} at the interface and surface could not be followed even in a qualitative way. Therefore the use of a Co marker in detecting the Fe magnetization profile in a one-to-one manner seems questionable.

Chapter 6

Summary

The results for orbital magnetic moments and the magnetic anisotropy presented in this thesis can be seen as a direct consequence of the spontaneous magnetization of the solid and the existence of spin-orbit coupling. We presented magnetic properties of various layered systems (metal surfaces and semiconductor/ferromagnet heterogeneous systems). In Chap. 2 the basic formalism underlying the results and applications was discussed. In the first part, density functional theory was introduced as a theoretical tool that provides a general framework for the calculation of the ground state properties using the charge density as the basic variable. In the second part, multiple scattering theory as a basis to solve the fully-relativistic Dirac-Kohn-Sham equations for the solid was introduced. The tight-binding version of the KKR method provided accurate means of investigating the energetics, magnetic and electronic structure of layered solid materials. In the third part, the contributions to the magnetic anisotropy energy were discussed and the connection between the magnetic anisotropy and the anisotropy of the orbital moment was introduced.

As shown in Chap. 3, in bcc structured Fe surface systems (the Fe crystal surfaces and Fe clusters), both spin and orbital magnetic moments are enhanced for atoms close to the surface. Their magnetization profiles exhibit an oscillatory structure, i.e. Friedel-like oscillations. There are common trends in the magnetization profiles of crystal surfaces and of free clusters. However, the surface enhancement as well as Friedel-like oscillations are more pronounced for clusters than for semi-infinite crystals. Spin magnetic moments at crystal surfaces and in clusters depend linearly on the effective coordination number and on the valence charge. A semi-empirical relationship between the magnetic moment and the effective coordination number could allow to account for some features of the measured dependence of the magnetic moment of free clusters on the cluster size. Concentrating on the crystal surfaces, the dipole-dipole (shape anisotropy) and the electronic contribution to the MAE has been calculated. The spontaneous magnetization direction in the film plane follows from the dominating shape anisotropy over the electronic MAE for all Fe crystal surfaces.

The electronic and magnetic properties of FePt surfaces in the CuAu structure were discussed in Chap. 4. As the bulk FePt is a strong ferromagnet and the spin moment of Fe in FePt is already rather high, the expected enhancement at the surface is only weak. The layer by layer decomposition of the MAE showed that the huge bulk contribution from the center of all FePt surface systems overrides the surface contribution and the resulting spontaneous magnetization is always perpendicular to the layers of the CuAu structure. It was also shown that the compositional disorder strongly destroys the MAE of the FePt alloy.

In Chap. 5 we presented a detailed study of the magnetic anisotropy of thin Fe films on (001) oriented GaAs. In agreement with experiment the films with more than 2 Fe layers were found to have an in-plane anisotropy with the easy axis along the [110]-direction. From an analysis of our theoretical results as well as from the experimental findings one can conclude that this anisotropy is caused primarily from the anisotropic chemical bonding to the GaAs substrate. Here, the layer by layer decomposition of the MAE brought the main argument for the assignment of the uniaxial in-plane MAE to the vicinity of the GaAs–Fe interface.

In addition, the influence of Au protective layers on magnetic properties of GaAs/Fe was shown and discussed. Using Brillouin light scattering technique in the Group of High Resolution Optical Spectroscopy and related Techniques (GHOST) of Perugia University (the group of Prof. G. Carlotti), a strong suppression of the in-plane anisotropy of Fe/GaAs films was found, especially its uniaxial component. By our theoretical study it was possible to demonstrate the decrease of the uniaxial anisotropy induced by Au, in qualitative agreement with the experiment. Moreover, the contributions of MAE could be assigned to the corresponding atoms and a deeper insight into the magnetic properties of the studied materials could be achieved.

The magnetization profile of Fe layers grown on GaAs(001) with atomic layer resolution from the interface up to the surface was detected by Giovanelli *et al.* [6–8]. It was achieved by measuring XMCD signals from Co impurities acting as magnetic marker. A complementary theoretical study was presented in Sec. 5.6. Results can be summarized as follows: the Co magnetic signal persists even for atoms in direct contact with the substrate ruling out the presence of a magnetically dead layer. The spin magnetic moment is reduced when approaching the interface and at the film surface. To achieve the reduction of the Co spin moment at the surface in theory, the diffusion of the As atoms to the Fe film surface had to be included. An increase of the orbital moment of Co when approaching the interface was observed. In addition, the feasibility of the Co marker in detecting actually the magnetization of the Fe atoms in GaAs/Fe system was discussed. As the spin magnetization profile of the Fe atoms in GaAs/Fe could not be followed by the Co marker even in a qualitative way, the use of a Co marker seems questionable.

Bibliography

[1] J. H. van Vleck, Phys. Rev. **52**, 1178 (1937).

[2] M. B. Stearns, *Magnetic Properties of 3d, 4d and 5d Elements, Alloys and Compounds* (Springer, Berlin, 1987), vol. III/19a of *Landolt-Börnstein, New Series*.

[3] D. Weller, G. R. Harp, R. F. C. Farrow, A. Cebollada, and J. Sticht, Phys. Rev. Letters **72**, 2097 (1994).

[4] E. S. Murdock, R. F. Simmons, and R. Davidson, IEEE Trans. Magn. **28**, 3078 (1992).

[5] G. A. Prinz, Science **282**, 1660 (1998).

[6] L. Giovanelli, C. S. Tian, P. L. Gastelois, G. Panaccione, M. Fabrizioli, M. Hochstrasser, M. Galaktionov, C. H. Back, and G. Rossi, Physica B **345**, 177 (2004).

[7] L. Giovanelli, G. Panaccione, G. Rossi, M. Fabrizioli, C. S. Tian, P. L. Gastelois, J. Fujii, and C. H. Back, Appl. Physics Lett. **87**, 42506 (2005).

[8] L. Giovanelli, G. Panaccione, G. Rossi, M. Fabrizioli, C. S. Tian, P. L. Gastelois, J. Fujii, and C. H. Back, Phys. Rev. B **72**, 45221 (2005).

[9] P. Hohenberg and W. Kohn, Phys. Rev. **136**, B 864 (1964).

[10] S. H. Vosko, L. Wilk, and M. Nusair, Can. J. Phys. **58**, 1200 (1980).

[11] V. Ozoliņš and M. Körling, Phys. Rev. B **48**, 18304 (1993).

[12] M. Battocletti, H. Ebert, and H. Akai, Phys. Rev. B **53**, 9776 (1996).

[13] I. Galanakis, S. Ostanin, M. Alouani, H. Dreysse, and J. Wills, Phys. Rev. B **61**, 599 (2000).

[14] R. Hafner, D. Spišák, R. Lorenz, and J. Hafner, Phys. Rev. B **65**, 184432 (2002).

[15] W. Kohn and L. J. Sham, Phys. Rev. **140**, A 1133 (1965).

[16] O. Gunnarson and B. I. Lundqvist, Phys. Rev. B **13**, 4274 (1976).

[17] U. von Barth and L. Hedin, J. Phys. C: Solid State Phys. **5**, 1629 (1972).

[18] P. Strange, *Relativistic Quantum Mechanics* (Cambridge University Press, Cambridge, 1998).

[19] A. K. Rajagopal and J. Callaway, Phys. Rev. B **7**, 1912 (1973).

[20] A. H. MacDonald and S. H. Vosko, J. Phys. C: Solid State Phys. **12**, 2977 (1979).

[21] M. V. Ramana and A. K. Rajagopal, Adv. Chem. Phys. **54**, 231 (1983).

[22] M. V. Ramana and A. K. Rajagopal, J. Phys. C: Solid State Phys. **12**, L845 (1979).

[23] T. H. Dupree, Ann. Phys. (New York) **15**, 63 (1961).

[24] J. L. Beeby, Proc. Roy. Soc. (London) A **302**, 113 (1967).

[25] N. A. W. Holzwarth, Phys. Rev. B **11**, 3718 (1975).

[26] J. Korringa, Physica **XIII**, 392 (1947).

[27] W. Kohn and N. Rostoker, Phys. Rev. **94**, 1111 (1954).

[28] J. S. Faulkner, J. Phys. C: Solid State Phys. **10**, 4661 (1977).

[29] J. S. Faulkner, Phys. Rev. B **19**, 6186 (1979).

[30] J. S. Faulkner and G. M. Stocks, Phys. Rev. B **21**, 3222 (1980).

[31] B. L. Gyorffy and G. M. Stocks, *Electrons in disordered Metals and at Metallic Surfaces* (Plenum Press, New York, 1979), p. 89.

[32] Y. Onodera and M. Okazaki, J. Phys. Soc. Japan **21**, 1273 (1966).

[33] P. Strange, H. Ebert, J. B. Staunton, and B. L. Gyorffy, J. Phys.: Condensed Matter **1**, 2959 (1989).

[34] P. Weinberger, *Electron Scattering Theory for Ordered and Disordered Matter* (Oxford University Press, Oxford, 1990).

[35] H. Ebert, in *Electronic Structure and Physical Properties of Solids*, edited by H. Dreyssé (Springer, Berlin, 2000), vol. 535 of *Lecture Notes in Physics*, p. 191.

[36] R. Zeller and P. H. Dederichs, Phys. Rev. Letters **42**, 1713 (1979).

[37] I. Turek, V. Drchal, J. Kudrnovský, M. Sob, and P. Weinberger, *Electronic structure of disordered alloys, surfaces and interfaces* (Kluwer Academic Publ., Boston, 1997).

[38] P. Soven, Phys. Rev. **156**, 809 (1967).

[39] P. J. Durham, B. L. Gyorffy, and A. J. Pindor, J. Phys. F: Met. Phys. **10**, 661 (1980).

[40] E. N. Economou, *Green's Functions in Quantum Physics* (Springer-Verlag, New York, 1990).

[41] B. L. Gyorffy and M. J. Stott, *Band Structure Spectroscopy of Metals and Alloys* (Academic Press, New York, 1973), p. 385.

[42] T. Huhne and H. Ebert, Phys. Rev. B **65**, 205125 (2002).

[43] M. E. Rose, *Relativistic Electron Theory* (Wiley, New York, 1961).

[44] H. Ebert and B. L. Gyorffy, J. Phys. F: Met. Phys. **18**, 451 (1988).

[45] L. Nordheim, Ann. Phys. (Leipzig) **9**, 607 (1931).

[46] J. Korringa, Ann. Phys. (Leipzig) **7**, 252 (1958).

[47] J. S. Faulkner, Prog. Mater. Sci. **27**, 3 (1982).

[48] H. Ebert and H. Akai, Mat. Res. Soc. Symp. Proc. **253**, 329 (1992).

[49] O. K. Andersen and O. Jepsen, Phys. Rev. Letters **53**, 2571 (1984).

[50] O. K. Andersen, A. V. Postnikov, and S. Y. Savrasov, Mat. Res. Soc. Symp. Proc. **253** (1992).

[51] L. Szunyogh, B. Újfalussy, P. Weinberger, and J. Kollar, Phys. Rev. B **49**, 2721 (1994).

[52] L. Szunyogh, B. Újfalussy, and P. Weinberger, Phys. Rev. B **51**, 9552 (1995).

[53] R. Zeller, P. H. Dederichs, B. Újfalussy, L. Szunyogh, and P. Weinberger, Phys. Rev. B **52**, 8807 (1995).

[54] K. Wildberger, *Tight-Binding-Korringa-Kohn-Rostoker-Methode und Grenzflächenreflektivität in magnetischen Systemen*, Ph.D. thesis, Jülich (1997).

[55] P. J. Braspenning and A. Lodder, Phys. Rev. B **49**, 10222 (1994).

[56] E. M. Godfrin, J. Phys.: Condensed Matter **3**, 7843 (1991).

[57] M. Brooks, Phys. Rev. **58**, 909 (1940).

[58] G. C. Fletcher, Proc. Roy. Soc. (London) A **67**, 505 (1954).

[59] M. S. S. Brooks, B. Johansson, O. Eriksson, and H. L. Skriver, Physica B **144**, 1 (1986).

[60] L. Fritsche, J. Noffke, and H. Eckardt, J. Phys. F: Met. Phys. **17**, 943 (1987).

[61] P. Strange, H. Ebert, J. B. Staunton, and B. L. Gyorffy, J. Phys.: Condensed Matter **1**, 3947 (1989).

[62] G. H. O. Daalderop, P. J. Kelly, and M. F. H. Schuurmans, Phys. Rev. B **41**, 11919 (1990).

[63] G. H. O. Daalderop, P. J. Kelly, and M. F. H. Schuurmans, Phys. Rev. B **44**, 12054 (1991).

[64] G. H. O. Daalderop, *Magneto-optical recording media based on Co/Pt-multilayers*, Ph.D. thesis, Technical University Delft (1991).

[65] G. Y. Guo, W. M. Temmerman, and H. Ebert, Physica B **172**, 61 (1991).

[66] J. Trygg, B. Johansson, O. Eriksson, and J. M. Wills, Phys. Rev. Letters **75**, 2871 (1995).

[67] S. V. Halilov, A. Y. Perlov, P. M. Oppeneer, A. N. Yaresko, and V. N. Antonov, Phys. Rev. B **57**, 9557 (1998).

[68] I. V. Solovyev, P. H. Dederichs, and I. Mertig, Phys. Rev. B **52**, 13419 (1995).

[69] S. S. A. Razee, J. B. Staunton, and F. J. Pinski, Phys. Rev. B **56**, 8082 (1997).

[70] S. S. A. Razee, J. B. Staunton, F. J. Pinski, B. Ginatempo, and E. Bruno, J. Appl. Physics **83**, 7097 (1998).

[71] J. L. Gay and R. Richter, Phys. Rev. Letters **56**, 2728 (1986).

[72] J. L. Gay and R. Richter, J. Appl. Physics **61**, 3362 (1987).

[73] G. H. O. Daalderop, P. J. Kelly, and M. F. H. Schuurmans, Phys. Rev. B **42**, 7270 (1990).

[74] G. Y. Guo, W. M. Temmerman, and H. Ebert, J. Phys.: Condensed Matter **3**, 8205 (1991).

[75] G. Y. Guo, W. M. Temmerman, and H. Ebert, J. Magn. Magn. Materials **104-107**, 1772 (1992).

[76] D. Wang, R. Wu, and A. J. Freeman, Phys. Rev. Letters **70**, 869 (1993).

[77] B. Újfalussy, L. Szunyogh, and P. Weinberger, *Properties of Complex Inorganic Solids* (Plenum Press, New York, 1997), p. 181.

[78] I. Cabria, H. Ebert, and A. Y. Perlov, Europhys. Lett. **51**, 209 (2000).

[79] I. Cabria, A. Perlov, and H. Ebert, Phys. Rev. B **63**, 104424/1 (2001).

[80] E. Sjöstedt, L. Nordström, F. Gustavsson, and O. Eriksson, Phys. Rev. Letters **89**, 267203 (2002).

[81] M. Kosuth, V. Popescu, H. Ebert, and G. Bayreuther, Europhys. Lett. **72**, 816 (2005).

[82] A. R. Mackintosh and O. K. Andersen, *Electrons at the Fermi Surface* (Cambridge University Press, Cambridge, 1980), chap. The Electronic Structure of Transition Metals.

[83] M. Weinert, R. E. Watson, and J. W. Davenport, Phys. Rev. B **32**, 2115 (1985).

[84] P. Bruno, Phys. Rev. B **39**, 865 (1989).

[85] G. van der Laan, J. Phys.: Condensed Matter **10**, 3239 (1998).

[86] L. Néel, J. Physique Le Radium **15**, 376 (1954).

[87] M. Brockmann, S. Miethaner, R. Onderka, M. Köhler, F. Himmelhuber, H. Regensburger, F. Bensch, T. Schweinböck, and G. Bayreuther, J. Appl. Physics **81**, 5047 (1997).

[88] G. Bayreuther, M. Dumm, B. Uhl, R. Meier, and W. Kipferl, J. Appl. Physics **93**, 8230 (2003).

[89] S. Ohnishi, A. J. Freeman, and M. Weinert, Phys. Rev. B **28**, 6741 (1983).

[90] C. L. Fu and A. J. Freeman, J. Magn. Magn. Materials **69**, L1 (1987).

[91] O. Hjortstam, J. Trygg, J. M. Wills, B. Johansson, and O. Eriksson, Phys. Rev. B **53**, 9204 (1996).

[92] D. Spišák and J. Hafner, J. Magn. Magn. Materials **168**, 257 (1997).

[93] A. M. N. Niklasson, B. Johansson, and H. L. Skriver, Phys. Rev. B **59**, 6373 (1999).

[94] J. Izquierdo, A. Vega, L. C. Balbás, D. Sánchez-Portal, J. Junquera, E. Artacho, J. M. Soler, and P. Ordejón, Phys. Rev. B **61**, 13639 (2000).

[95] V. Popescu, H. Ebert, B. Nonas, and P. H. Dederichs, Phys. Rev. B **64**, 184407 (2001).

[96] J. L. Rodriguez-López, J. Dorantes-Dávila, and G. M. Pastor, Phys. Rev. B **57**, 1040 (1998).

[97] O. Eriksson, G. W. Fernando, A. M. Boring, and R. C. Albers, Solid State Commun. **78**, 801 (1991).

[98] O. Sipr, M. Kosuth, and H. Ebert, Surf. Sci. **566-568**, 268 (2004).

[99] O. Sipr, M. Kosuth, and H. Ebert, J. Magn. Magn. Materials **272-276**, 713 (2004).

[100] O. Sipr, M. Kosuth, and H. Ebert, Phys. Rev. B **70**, 174423 (2004).

[101] G. M. Pastor, J. Dorantes-Dávila, and K. H. Bennemann, Phys. Rev. B **40**, 7642 (1989).

[102] A. Vega, J. Dorantes-Dávila, L. C. Balbás, and G. M. Pastor, Phys. Rev. B **47**, 4742 (1993).

[103] J. A. Franco, A. Vega, and F. Aguilera-Granja, Phys. Rev. B **60**, 434 (1999).

[104] D. R. Salahub and R. P. Messmer, Surf. Sci. **106**, 415 (1981).

[105] C. Y. Yang, K. H. Johnson, D. R. Salahub, J. Kaspar, and R. P. Messmer, Phys. Rev. B **24**, 5673 (1981).

[106] K. Lee, J. Callaway, and S. Dhar, Phys. Rev. B **30**, 1724 (1984).

[107] O. Diéguez, M. M. G. Alemany, C. Rey, P. Ordejón, and L. J. Gallego, Phys. Rev. B **63**, 205407 (2001).

[108] A. V. Postnikov, P. Entel, and J. M. Soler, Eur. Phys. J. D **25**, 261 (2003).

[109] R. A. Guirado-López, J. Dorantes-Dávila, and G. M. Pastor, Phys. Rev. Letters **90**, 226402 (2003).

[110] N. Papanikolaou, B. Nonas, S. Heinze, R. Zeller, and P. H. Dederichs, Phys. Rev. B **62**, 11118 (2000).

[111] M. S. S. Brooks, Physica B **130**, 6 (1985).

[112] O. Eriksson, B. Johansson, R. C. Albers, A. M. Boring, and M. S. S. Brooks, Phys. Rev. B **42**, 2707 (1990).

[113] H. Ebert and M. Battocletti, Solid State Commun. **98**, 785 (1996).

[114] V. I. Anisimov, F. Aryasetiawan, and A. I. Lichtenstein, J. Phys.: Condensed Matter **9**, 767 (1997).

[115] H. Ebert, A. Perlov, and S. Mankovsky, Solid State Commun. **127**, 443 (2003).

[116] P. Bruno, in *IFF-Ferienkurs*, edited by F. Jülich (1993), p. 24.1.

[117] P. Ballone and R. O. Jones, Chem. Phys. Lett. **233**, 632 (1995).

[118] A. N. Andriotis and M. Menon, Phys. Rev. B **57**, 10069 (1998).

[119] H. Ebert, *The munich* SPR-KKR *package, version 2.1.1*, http://olymp.cup.uni-muenchen.de/ak/ebert/SPRKKR (2002).

[120] M. Cook and D. A. Case, *computer code* XASCF, Quantum Chemistry Program Exchange (1980).

[121] O. Šipr and A. Šimůnek, J. Phys.: Condensed Matter **13**, 8519 (2001).

[122] O. Šipr and H. Ebert, Czech. J. Phys. **53**, 55 (2003).

[123] T. Oda, A. Pasquarello, and R. Car, Phys. Rev. Letters **80**, 3622 (1998).

[124] T. Oda, A. Pasquarello, and R. Car, Phys. Rev. B **62**, 11556 (2000).

[125] G. Kresse and J. Hafner, Phys. Rev. B **47**, 558 (1993).

[126] G. Kresse and J. Furthmüller, Phys. Rev. B **54**, 11169 (1996).

[127] J. Minár, private communication (2004).

[128] A. P. Cracknell, J. Phys. C: Solid State Phys. **2**, 1425 (1969).

[129] G. M. Pastor, J. Dorantes-Dávila, Š. Pick, and H. Dreyssé, Phys. Rev. Letters **75**, 326 (1995).

[130] K. W. Edmonds, C. Binns, S. H. Baker, S. C. Thornton, C. Norris, J. B. Goedkoop, M. Finazzi, and N. B. Brookes, Phys. Rev. B **60**, 472 (1999).

[131] K. W. Edmonds, C. Binns, S. H. Baker, M. J. Maher, S. C. Thornton, O. Tjernberg, and N. B. Brookes, J. Magn. Magn. Materials **231**, 113 (2001).

[132] P. Ohresser, G. Ghiringhelli, O. Tjernberg, N. B. Brookes, and M. Finazzi, Phys. Rev. B **62**, 5803 (2000).

[133] P. Gambardella, S. Rusponi, M. Veronese, S. S. Dhesi, C. Grazioli, A. Dallmeyer, I. Cabria, R. Zeller, P. H. Dederichs, K. Kern, C. Carbone, and H. Brune, Science **300**, 1130 (2003).

[134] P. Carra, B. T. Thole, M. Altarelli, and X. Wang, Phys. Rev. Letters **70**, 694 (1993).

[135] C. Ederer, M. Komelj, and M. Fähnle, Phys. Rev. B **68**, 052402 (2003).

[136] L. M. Falicov and G. A. Somorjai, Proc. Natl. Acad. Sci. USA **82**, 2207 (1985).

[137] A. J. Freeman and R. Q. Wu, J. Magn. Magn. Materials **100**, 497 (1991).

[138] F. Liu, M. R. Press, S. N. Khanna, and P. Jena, Phys. Rev. B **39**, 6914 (1989).

[139] D. Tománek, S. Mukherjee, and K. H. Bennemann, Phys. Rev. B **28**, 665 (1983).

[140] G. M. Pastor, J. Dorantes-Dávila, and K. H. Bennemann, Chem. Phys. Lett. **148**, 459 (1988).

[141] J. Zhao, X. Chen, Q. Sun, F. Liu, and G. Wang, Phys. Lett. A **205**, 308 (1995).

[142] F. Aguilera-Granja, J. M. Montejano-Carrizales, and J. L. Morán-López, Phys. Lett. A **242**, 255 (1998).

[143] T. B. Massalski, ed., *Binary Phase Diagrams* (American Society for Metals, Metals Park, OH, 44073).

[144] D. Weller and A. Moser, IEEE Trans. Magn. **35**, 4423 (1999).

[145] K. R. Coffey, M. A. Parker, and J. K. Howard, IEEE Trans. Magn. **31**, 2737 (1995).

[146] C. P. Luo, IEEE Trans. Magn. **31**, 2764 (1995).

[147] D. Weller, *Spin-orbit influenced spectroscopies of magnetic solids* (Springer, Berlin, 1996), vol. 466 of *Lecture Notes in Physics*, p. 1.

[148] S. H. Sun, Science **287**, 1989 (2000).

[149] J. A. Christodoulides, Y. Huang, Y. Zhang, G. C. Hadjipanayis, I. Panagiotopoulos, and D. Niarchos, J. Appl. Physics **87**, 6938 (2000).

[150] M. M. Schwickert, K. A. Hannibal, M. F. Toney, M. Best, L. Folks, J. U. Thiele, A. J. Kellock, and D. Weller, J. Appl. Physics **87**, 6956 (2000).

[151] C. P. Luo, S. H. Liou, and D. J. Sellmyer, J. Appl. Physics **87**, 6941 (2000).

[152] S. Y. Bae, K. H. Shin, J. Y. Jeong, and J. G. Kim, J. Appl. Physics **87**, 6953 (2000).

[153] I. Galanakis, M. Alouani, and H. Dreysse, Phys. Rev. B **62**, 6475 (2005).

[154] R. F. C. Farrow, D. Weller, R. F. Marks, M. F. Toney, A. Cebollada, and G. R. Harp, J. Appl. Physics **79**, 5967 (1996).

[155] J. U. Thiele, L. Folks, M. F. Toney, and D. K. Weller, J. Appl. Physics **84**, 5686 (1996).

[156] P. Ravindran, A. Kjekshus, H. Fjellvåg, P. James, L. Nordström, B. Johansson, and O. Eriksson, Phys. Rev. B **63**, 144409 (2001).

[157] S. Ostanin, S. S. A. Razee, J. B. Staunton, B. Ginatempo, and E. Bruno, J. Appl. Physics **93**, 453 (2003).

[158] J. B. Staunton, S. Ostanin, S. S. A. Razee, B. L. Gyorffy, L. Szunyogh, B. Ginatempo, and E. Bruno, Phys. Rev. Letters **93**, 257204 (2004).

[159] A. Perlov, H. Ebert, A. N. Yaresko, V. N. Antonov, and D. Weller, Solid State Commun. **105**, 273 (1998).

[160] M. N. Baibich, J. M. Broto, A. Fert, F. N. V. Dau, F. Petroff, P. Etienne, G. Creuzet, A. Friederich, and J. Chazelas, Phys. Rev. Letters **61**, 2472 (1988).

[161] M. Jullière, Phys. Letters **54A**, 225 (1975).

[162] J. S. Moodera, L. R. Kinder, T. M. Wong, and R. Meservey, Phys. Rev. Letters **74**, 3273 (1995).

[163] H. J. Zhu, M. Ramsteiner, H. Kostial, M. Wassermeier, H. P. Schonherr, and K. H. Ploog, Phys. Rev. Letters **87**, 016601/1 (2001).

[164] M. Zenger, J. Moser, W. Wegscheider, D. Weiss, and T. Dietl, J. Appl. Physics **96**, 2400 (2004).

[165] J. Moser, M. Zenger, C. Gerl, D. Schuh, R. Meier, P. Chen, G. Bayreuther, W. Wegscheider, D. Weiss, C.-H. Lai, R.-T. Huang, M. Kosuth, *et al.*, Appl. Physics Lett. **89**, 162106 (20006).

[166] W. E. Pickett and D. A. Papaconstantopoulos, Phys. Rev. B **34**, 8372 (1986).

[167] W. H. Butler, X. G. Zhang, X. Wang, J. van Ek, and J. M. MacLaren, J. Appl. Physics **81**, 5518 (1997).

[168] J. M. MacLaren, W. H. Butler, and X. G. Zhang, J. Appl. Physics **83**, 6521 (1998).

[169] W. E. Pickett and D. A. Papaconstantopoulos, Phys. Rev. B **34**, 8372 (1986).

[170] W. E. Pickett and D. A. Papaconstantopoulos, J. Appl. Physics **61**, 3735 (1987).

[171] A. Continenza, S. Massidda, and A. J. Freeman, Phys. Rev. B **42**, 2904 (1990).

[172] M. Freyss, N. Papanikolaou, V. Bellini, R. Zeller, and P. H. Dederichs, Phys. Rev. B **66**, 014445/1 (2002).

[173] B. Sanyal and S. Mirbt, Phys. Rev. B **65**, 144435/1 (2002).

[174] M. Kosuth, J. Minar, I. Cabria, A. Perlov, V. Crisan, H. Ebert, and H. Akai, Phase Transitions **75**, 113 (2002).

[175] J. M. MacLaren, X. G. Zhang, W. H. Butler, and X. Wang, Phys. Rev. B **59**, 5470 (1999).

[176] P. Mavropoulos, N. Papanikolaou, and P. H. Dederichs, Phys. Rev. Letters **85**, 1088 (2000).

[177] W. H. Butler, X. G. Zhang, T. C. Schulthess, and J. M. MacLaren, Phys. Rev. B **63**, 054416/1 (2001).

[178] H. C. Herper, P. Weinberger, A. Vernes, L. Szunyogh, and C. Sommers, Phys. Rev. B **64**, 184442/1 (2001).

[179] P. Weinberger, L. Szunyogh, C. Blaas, and C. Sommers, Phys. Rev. B **64**, 184429/1 (2001).

[180] M. Freyss, N. Papanikolaou, V. Bellini, R. Zeller, P. H. Dederlchs, and I. Turek, J. Magn. Magn. Materials **240**, 117 (2002).

[181] M. Freyss, P. Mavropoulos, N. Papanikolaou, V. Bellini, R. Zeller, and P. H. Dederichs, Phase Transitions **75**, 159 (2002).

[182] O. Wunnicke, P. Mavropoulos, and P. H. Dederichs, J. of Superconductivity **16**, 171 (2003).

[183] V. Popescu, H. Ebert, N. Papanikolaou, P. H. Dederichs, and R. Zeller, J. Phys.: Condensed Matter **16**, S5579 (2004).

[184] V. Popescu, H. Ebert, N. Papanikolaou, P. H. Dederichs, and R. Zeller, Phys. Rev. B **72**, 184427 (2005).

[185] F. Gustavsson, E. Nordström, V. H. Etgens, M. Eddrief, E. Sjöstedt, R. Wappling, and J. M. George, Phys. Rev. B **66**, 024405/1 (2002).

[186] J. J. Krebs, B. T. Jonker, and G. A. Prinz, J. Appl. Physics **61**, 3744 (1987).

[187] R. Meckenstock, D. Spoddig, K. Himmelbauer, H. Krenn, M. Doi, W. Keune, Z. Frait, and J. Pelzl, J. Magn. Magn. Materials **240**, 410 (2002).

[188] M. Marangolo, F. Gustavsson, M. Eddrief, P. Sainctavit, V. H. Etgens, V. Cros, F. Petroff, J. M. George, P. Bencok, and N. B. Brookes, Phys. Rev. Letters **88**, 217202/1 (2002).

[189] E. Reiger, E. Reinwald, G. Garreau, M. Ernst, M. Zölfl, F. Bensch, S. Bauer, H. Preis, and G. Bayreuther, J. Appl. Physics **87**, 5923 (2000).

[190] S. A. Haque, A. Matsuo, Y. Yamamoto, and H. Hori, J. Magn. Magn. Materials **247**, 117 (2002).

[191] E. M. Kneedler, B. T. Jonker, P. M. Thibado, R. J. Wagner, B. V. Shanabrook, and L. J. Whitman, Phys. Rev. B **56**, 8163 (1997).

[192] Y. B. Xu, E. T. M. Kernohan, D. J. Freeland, A. Ercole, M. Tselepi, and J. A. C. Bland, Phys. Rev. B **58**, 890 (1998).

[193] M. Brockmann, M. Zölfl, S. Miethaner, and G. Bayreuther, J. Magn. Magn. Materials **198-199**, 384 (1999).

[194] G. W. Anderson, M. C. Hanf, P. R. Norton, M. Kowalewski, K. Myrtle, and B. Heinrich, J. Appl. Physics **79**, 4954 (1996).

[195] M. Dumm, F. Bensch, R. Moosbühler, M. Zölfl, M. Brockmann, and G. Bayreuther, Magnetic Storage Systems Beyond 2000 **41**, 555 (2002).

[196] F. Bensch, R. Moosbühler, and G. Bayreuther, J. Appl. Physics **91**, 8754 (2002).

[197] R. Urban, G. Woltersdorf, and B. Heinrich, Phys. Rev. Letters **87**, 217204/1 (2001).

[198] R. Moosbühler, F. Bensch, M. Dumm, and G. Bayreuther, J. Appl. Physics **91**, 8757 (2002).

[199] M. Madami, S. Tacchi, G. Carlotti, G. Gubbiotti, and R. L. Stamps, Phys. Rev. B **69**, 144408 (2004).

[200] K. Zakeri, T. Kebe, J. Lindner, and M. Farle, J. Magn. Magn. Materials **299**, L1 (2005).

[201] S. McPhail, C. M. Gurtler, F. Montaigne, Y. B. Xu, M. Tselepi, and J. A. C. Bland, Phys. Rev. B **67**, 24409 (2003).

[202] J. J. Krebs, B. T. Jonker, and G. A. Prinz, J. Appl. Physics **61**, 2596 (1987).

[203] M. Gester, C. Daboo, R. J. Hicken, S. j Gray, A. Ercole, and J. A. C. Bland, J. Appl. Physics **80**, 347 (1996).

[204] Y. B. Xu, D. J. Freeland, M. Tselepi, and J. A. C. Bland, J. Appl. Physics **87**, 6110 (2000).

[205] F. Bensch, G. Garreau, R. Moosbühler, G. Bayreuther, and E. Beaurepaire, J. Appl. Physics **89**, 7133 (2001).

[206] http://sham.phys.sci.osaka u.ac.jp/~kkr/ .

[207] K. Wildberger, R. Zeller, and P. H. Dederichs, Phys. Rev. B **55**, 10074 (1997).

[208] B. Lepine, S. Ababou, A. Guivarc'h, G. Jezequel, S. Deputier, R. Guerin, A. Filipe, A. Schuhl, F. Abel, C. Cohen, A. Rocher, and J. Crestou, J. Appl. Physics **83**, 3077 (1998).

[209] H. Katsuraki and N. Achiwa, J. Phys. Soc. Japan **21**, 2238 (1966).

[210] M. Yuzuri, R. Tahara, and Y. Nakamura, J. Phys. Soc. Japan **48**, 1937 (1980).

[211] M. W. Ruckman, J. J. Joyce, and J. H. Weaver, Phys. Rev. B **33**, 7029 (1986).

[212] S. A. Chambers, F. Xu, H. W. Chen, I. M. Vitomirov, S. B. Anderson, and J. H. Weaver, Phys. Rev. B **34**, 6605 (1986).

[213] M. Zölfl, M. Brockmann, M. Köhler, S. Kreuzer, T. Schweinböck, S. Miethaner, F. Bensch, and G. Bayreuther, J. Magn. Magn. Materials **175**, 16 (1997).

[214] J. W. Freeland, I. Coulthard, W. J. Antel, and A. P. J. Stampfl, Phys. Rev. B **63**, 193301/1 (2001).

[215] A. Filipe, A. Schuhl, and P. Galtier, Appl. Physics Lett. **70**, 129 (1997).

[216] A. Filipe and A. Schuhl, J. Appl. Physics **81**, 4359 (1997).

[217] M. Doi, B. R. Cuenya, W. Keune, T. Schmitte, A. Nefedov, H. Zabel, D. Spoddig, R. Meckenstock, and Pelz, J. Magn. Magn. Materials **240**, 407 (2002).

[218] F. Wilhelm, P. Poulopoulos, G. Ceballos, H. Wende, K. Baberschke, P. Srivastava, D. Benea, H. Ebert, M. Angelakeris, N. K. Flevaris, D. Niarchos, A. Rogalev, *et al.*, Phys. Rev. Letters **85**, 413 (2000).

[219] H. A. Dürr, E. Dudzik, S. S. Dhesi, J. B. Goedkoop, G. van der Laan, M. Belakhovsky, C. Mocuta, A. Marty, and Y. Samson, Science **284**, 2166 (1999).

[220] N. Ishimatsu, H. Hashizume, S. Hamada, N. Hosoito, C. Nelson, C. Venkataraman, G. Srajer, and J. Lang, Phys. Rev. B **60**(13), 9596 (1999).

[221] B. T. Thole, P. Carra, F. Sette, and G. van der Laan, Phys. Rev. Letters **68**, 1943 (1992).

[222] G. Schütz, W. Wagner, W. Wilhelm, P. Kienle, R. Zeller, R. Frahm, and C. Materlik, Phys. Rev. Letters **58**, 737 (1987).

[223] H. Ebert, Rep. Prog. Phys. **59**, 1665 (1996).

[224] H. Ebert, V. Popescu, and D. Ahlers, Phys. Rev. B **60**, 7156 (1999).

[225] M. G. Samant, J. Stöhr, S. S. Parkin, G. A. Held, B. D. Hermsmeier, F. Herman, M. van Schilfgaarde, L.-C. Duda, D. C. Mancini, N. Waasdahl, and R. Nakajima, Phys. Rev. Letters **72**, 1112 (1994).

[226] G. van der Laan, Phys. Rev. Letters **82**, 640 (1999).

[227] G. Woltersdorf, private communication (2005).

[228] S. T. Purcell, B. Heinrich, and A. S. Arrott, J. Appl. Physics **64**, 5337 (1988).

[229] E. Kneedler, P. M. Thibado, B. T. Jonker, B. R. Bennett, B. V. Shanabrook, R. J. Wagner, and L. J. Whitman, J. Vac. Sci. Technol. B **14**, 3193 (1996).

[230] S. Mirbt, B. Sanyal, C. Isheden, and B. Johansson, Phys. Rev. B **67**, 155421 (2003).

[231] B. Hillebrands, *Light scattering in Solids VII* (Springer Series Topics Applied Physics, Springer-Verlag, Berlin, 2000).

[232] G. Carlotti and G. Gubbiotti, La Rivista del Nuovo Cimento **22**, 1 (1999).

[233] J. F. Cochran, *Ultrathin Magnetic Structure II* (Springer-Verlag, Berlin, 1994).

[234] B. Hillebrands and G. Güntherodt, *Ultrathin Magnetic Structure II* (Springer-Verlag, Berlin, 1994).

[235] B. Hillebrands, P. Baumgart, and G. Güntherodt, Phys. Rev. B **36**, 2450 (1988).

[236] P. Baumgart, B. Hillebrands, and G. Güntherodt, J. Magn. Magn. Materials **93**, 225 (1991).

[237] S. Scheurer, R. Allenspach, P. Xhonneux, and E. Courtens, Phys. Rev. B **48**, 9890 (1993).

[238] R. J. Hicken, A. Ercole, S. J. Gray, C. Daboo, and J. A. C. Bland, J. Appl. Physics **79**, 4987 (1996).

[239] R. J. Hicken, S. J. Gray, A. Ercole, C. Daboo, D. J. Freeland, E. Gu, E. Ahmad, and J. A. C. Bland, Phys. Rev. B **55**, 5898 (1997).

[240] L. Albini, G. Carlotti, G. Gubbiotti, M. Madami, and S. Tacchi, J. Appl. Physics **89**, 7383 (2001).

[241] W. Damon and J. R. Eshbach, J. Phys. Chem. Solids **19**, 308 (1961).

[242] R. E. Camley, P. Grunberg, and C. M. Mayr, Phys. Rev. B **26**, 2609 (1982).

[243] G. Gubbiotti, G. Carlotti, and B. Hillebrands, J. Phys.: Condensed Matter **10**, 2171 (1998).

[244] D. Liljequist, M. Ismail, K. Saneyoshi, K. Debusmann, W. Keune, R. A. Brand, and W. Kiauka, Phys. Rev. B **31**, 4137 (1985).

[245] M. Dumm, B. Uhl, M. Zölfl, W. Kipferl, and G. Bayreuther, J. Appl. Physics **91**, 8763 (2002).

[246] M. Liberati, G. Panaccione, F. Sirotti, P. Prieto, and G. Rossi, Phys. Rev. B **59**, 4201 (1999).

[247] A. Ankudinov and J. J. Rehr, Phys. Rev. B **51**, 1282 (1995).

[248] J. Minár, S. Bornemann, O. Šipr, S. Polesya, and H. Ebert, Appl. Physics A **82**, 139 (2005).

[249] E. D. Fabrizio, G. Mazzone, C. Petrillo, and F. Sacchetti, Phys. Rev. B **40**, 9502 (1989).

[250] K. Schwarz, P. Mohn, P. Blaha, and J. Kübler, J. Phys. F: Met. Phys. **14**, 2659 (1984).

Acknowledgements

First of all I would like to thank to my supervisor Prof. Dr. H. Ebert for his scientific guidance. Many thank also to my colleagues Dr. D. Benea, S. Bornemann, Apl.Prof. Dr. J. Braun, S. Chadov, Prof. Dr. V. Crisan, Dr. M. Deng, Dr. H. Freyer, Dr. T. Huhne, M. Kardinal, Dr. D. Ködderitzsch, Dr. S. Mankovsky, Dr. J. Minár, Dr. A. Perlov, S. Polesya, Dr. V. Popescu and Dr. C. Zecha for a nice working atmosphere and for fruitful discussions.

My thesis could not have been completed without the rich discussions with various theoreticians and experimentalists: Prof. Dr. C. Back, Apl.Prof. Dr. G. Bayreuther, Dr. M. Fabrizioli, G. Kuhn, Dr. S. Ostanin, Dr. O. Šipr, Dr. S. Tacchi, Dr. G. Woltersdorf and Kh. Zakeri.

I am also indebted to Prof. Dr. H. Akai for his support during my stay in Japan.

I would like to thank to my parents, my brother and sister, and especially to my wife Gabriela.

Financial support from the DFG-Förderprojekt FOG 370/2-1 within the research group *Ferromagnet-Halbleiter-Nanostrukturen: Transport, elektrische und magnetische Eigenschaften* is also acknowledged.

Curriculum vitae – Lebenslauf

PERSÖNLICHE ANGABEN

- Name: Michal Košuth
- Geburtsdatum: 2 Juli 1976
- Geburtsort: Bratislava, Slowakische Republik
- Familienstand: verheiratet

SCHULBILDUNG

- 1982 – 1990: Grundschule in Bratislava, Slowakische Republik
- 1990 – 1994: Gymnasium Bilíkova in Bratislava, Slowakische Republik
- Juni 1994: Abschluss mit einem in Österreich anerkannten Abitur

STUDIUM

- 1994 – 1999: Slowakische Technische Universität, Chemisch-technologische Fakultät, Bratislava
- Januar 1999 – Juni 1999: Diplomarbeit in theoretischer Festkörperchemie. Titel: "Berechnung der Elektronenstruktur der modellierten Polysilane."
- Juni 1999: Diplomprüfung, Fach Physikalische Chemie

BERUF

- September 1999 – Februar 2000: externer Doktorand an der Chemisch-technologischen Fakultät der Slowakischen Technologischen Universität in Bratislava, Slowakei

- März 2000 – Juli 2006: wissenschaftlicher Angestellter und Doktorand am Institut für Physikalische Chemie an der Ludwig–Maximilians–Universität München.

List of Publications (1999-2006)

To a large extent, the original work within this thesis has been previously published in various scientific journals or presented at conferences and workshops. A comprehensive list of the publications and workshops is given below:

1. P. Pelikán, M. Košuth, S. Biskupič, J. Noga, M. Straka, A. Zajac, P. Baňacký – *Electron Structure of Polysilanes. Are These Polymers One-Dimensional Systems?* – International Journal of Quantum Chemistry **84**, 157 (2001)

2. M. Košuth, J. Minár, I. Cabria, A. Perlov, V. Crisan, H. Ebert, A. Akai – *Electronic and magnetic properties of ferromagnet-semiconductor heterostructure systems* – Phase Transitions **75**, 113 (2002)

3. M. Košuth, V. Popescu, J. Minár, A. Perlov, H. Ebert – *Electronic and magnetic properties of ferromagnet-semiconductor heterostructure systems* – Phase Transitions **76**, 501 (2003)

4. O. Šipr, M. Košuth, H. Ebert – *Ab initio calculation of magnetic structure of small iron nanoclusters* – J. Magn. Magn. Materials **272-276**, 713 (2004)

5. O. Šipr, M. Košuth, H. Ebert – *Magnetic structure of iron clusters and iron crystal surfaces* – Surface Science **566-568**, 268 (2004)

6. O. Šipr, M. Košuth, H. Ebert – *Magnetic structure of free iron clusters compared to iron crystal* – Phys. Rev. B **70**, 174423 (2004)

7. H. Ebert, S. Bornemann, J. Minár, M. Kosuth, O. Šipr, P. H. Dederichs, R. Zeller, I. Cabria – *Electronic and magnetic properties*

149

of free and supported transition metal clusters – Phase Transitions **78**, 71-83 (2005)

8. M. Kosuth, V. Popescu, H. Ebert, G. Bayreuther – *Magnetic anisotropy of thin Fe films on GaAs* – Europhys. Lett. **72**, 816 (2005)

9. J. Moser, M. Zenger, C. Gerl, D. Schuh, R. Meier, P. Chen, G. Bayreuther, W. Wegscheider, D. Weiss, C.-H. Lai, R.-T. Huang, M. Kosuth, H. Ebert – *Bias dependent inversion of tunneling magnetoresistance in Fe/GaAs/Fe tunnel junctions* – Appl. Physics Lett. **89**, 162106 (2006)

10. S. Tacchi, A. Stollo, G. Gubbiotti, G. Carlotti, M. Košuth, H. Ebert – *Influence of Au capping layer on the magnetic properties of Fe/GaAs(001) ultrathin films* – Surface Science **601**, 4311 (2007)

List of Presentations

1. M. Košuth, V. Crisan, H. Ebert – **poster**, Psi-k 2000 Conference – title: *Electronic properties of Fe on GaAs(100) surfaces*, Schwäbisch Gmünd, Germany, August 2000

2. M. Košuth, H. Ebert, H. Akai – **poster**, The European Graduate School on Condensed Matter – title: *Electronic and magnetic properties of Fe on GaAs(100) surfaces*, Prague, Czech Republic, June 2001

3. M. Košuth, H. Ebert, H. Akai – **talk**, European 'Research Training' Network (RTN) "Computational Magnetoelectronics" Workshop – title: *Electronic and magnetic properties of Fe on GaAs(100) surfaces*, Budapest, Hungary, September 2001

4. M. Košuth, H. Ebert, H. Akai – **talk**, Spring Meeting of DPG – title: *Electronic and magnetic properties of Fe/GaAs heterogeneous systems and related compounds*, Regensburg, Germany, March 2002

5. M. Košuth, H. Ebert, H. Akai – **poster**, 9th General Conference of the Condensed Matter Division of the European Physical Society – title: *Electronic and magnetic properties of Fe/GaAs heterogeneous systems and related compounds*, Brighton, UK, April 2002

6. M. Košuth, V. Popescu, H. Ebert, I. Cabria, J. Minár – **poster**, RTN "Computational Magnetoelectronics" Workshop – title: *Electronic and magnetic properties of ferromagnet/semiconductor interfaces*, Ile d'Oleron, France, September 2002

7. M. Košuth, J. Minár, H. Ebert – **invited talk**, Hands-on Course: KKR band structure and spectroscopy calculations – title: *The Munich SPRKKR-program package: Short introduction to Density Functional Theory (DFT) and Korringa-Kohn-Rostoker (KKR) method*, Argonne National Laboratory, USA, January 2003

8. M. Košuth, V. Popescu, H. Ebert – **talk**, Spring Meeting of DPG – title: *Electronic and magnetic properties of ferromagnet/semiconductor interfaces*, Dresden, Germany, March 2003

9. H. Ebert, V. Popescu, M. Košuth, A. Perlov, J. Minár, R. Zeller, P. H. Dederichs, N. Papanikolaou – **talk**, joint Workshop of ESF, NEDO network and RTN "Spin mesoscopics" – title: *Electronic, magnetic and transport properties of ferromagnet-semiconductor heterostructure systems*, University of Twente, The Netherlands, March 2003

10. M. Košuth, V. Popescu, H. Ebert – **talk**, Summer School on New Magnetics – title: *Electronic, magnetic and transport properties of ferromagnet-semiconductor heterostructure systems*, Bedlewo, Poland, September 2003

11. M. Košuth, V. Popescu, H. Ebert, R. Zeller, P. H. Dederichs – **poster**, RTN "Computational Magnetoelectronics" – title: *Electronic, magnetic and transport properties of ferromagnet-semiconductor heterostructure systems*, Max Planck Insitute Halle, Germany, October 2003

12. M. Košuth, V. Popescu, H. Ebert – **poster**, KKR-workshop: new developments, applications and collaborations – title: *Electronic, magnetic and transport properties of ferromagnet-semiconductor heterostructure systems*, University of Munich, Munich, Germany, February 2004

13. M. Košuth, V. Popescu, H. Ebert – **talk**, Spring Meeting of DPG – title: *Magnetism of thin Fe films on GaAs*, Regensburg, Germany, March 2004

14. M. Košuth, V. Popescu, H. Ebert – **poster**, "International Symposion on Structure and Dynamics of Heterogeneous Systems", SDHS 2004 – title: *Electronic, magnetic and transpot properties of ferromagnet-semiconductor heterostructure systems*, Duisburg, Germany, November 2004

15. M. Kosuth, V. Popescu, H. Ebert – **talk**, Spring Meeting of DPG – title: *Magnetism of thin Fe films on GaAs*, Berlin, Germany, March 2005

16. M. Kosuth, V. Popescu, H. Ebert – **poster**, Intl. Workshop on Spin Phenomena in Reduced Dimensions – title: *Magnetism of thin Fe films on GaAs*, Regensburg, Germany, March 2005

17. M. Košuth, V. Popescu, H. Ebert – **poster**, Psi-k 2005 Conference – Title: *Effect of Au covering layers on the magnetic anisotropy of thin Fe films on GaAs*, Schwäbisch Gmünd, Germany, September 2005

18. M. Kosuth, S. Polesya, V. Popescu, H. Ebert – **talk**, Spring Meeting of DPG – title: *Magnetism of ultrathin GaMnAs films on GaAs*, Dresden, Germany, March 2006

19. M. Košuth, S. Lowitzer, H. Ebert – **talk**, SFB Workshop Frauenchiemsee – title: *Magnetic anisotropy of layered systems*, Frauenchiemsee, Germany, 2006